LESSONS

IN

EXPERIMENTAL AND PRACTICAL GEOMETRY

LESSONS
IN
EXPERIMENTAL AND PRACTICAL GEOMETRY

BY

H. S. HALL, M.A.

AND

F. H. STEVENS M. A.

GYAN

Published by

Gyan Publishing House
5, Ansari Road
Daryaganj, New Delhi-110002
Phone: 011-47034999, 9811692060
E-mail: books@gyanbooks.com

Distribution Network
gyanbooks.com
India, USA, Canada, UK, Australia

ISBN: 978-93-6208-068-4 (HB)
First Published, 1906

2nd Impression 2024

Printed at: Gyan Press, Delhi.

LESSONS IN EXPERIMENTAL AND PRACTICAL GEOMETRY
Author: H. S. HALL

LESSONS IN EXPERIMENTAL AND PRACTICAL GEOMETRY

LESSONS

IN

EXPERIMENTAL AND PRACTICAL GEOMETRY

BY

H. S. HALL, M.A.

AND

F. H. STEVENS, M.A.

London

MACMILLAN AND CO., LIMITED

NEW YORK: THE MACMILLAN COMPANY

1906

PREFACE.

To give to a young pupil clear mental pictures should be the first object of geometrical teaching: to enable him to express geometrical ideas in the form and order required by strict deductive reasoning is a second and distinct object. Experience shews that these two aims may to some extent be separated with advantage; and accordingly Formal Geometry is now very generally preceded by a short preliminary course of practical and experimental work.

In the preface to our *School Geometry* it is suggested that a suitable introduction to that book would consist of "Easy Exercises in Drawing to illustrate Definitions; Measurements of Lines and Angles; The Use of Compasses and Protractor; Problems on Bisection, Parallels, Perpendiculars; The Use of Set Squares; and the Construction of Triangles and Quadrilaterals: these problems to be informally explained, and the results verified by measurement. Concurrently there should be Exercises in Drawing and Measurement designed to lead inductively to the more important Theorems of Part I." It is the purpose of these *Lessons* to supply such an introductory course.

To this scheme we have added very simple chapters on Areas, on Circles and Polygons, and on the Forms of some Solid Figures; but it is not intended that these Sections should necessarily be taken before demonstrative geometry is begun.

This experimental and constructive work should not be allowed to keep a pupil back. He may probably be put to it six months or a year before a start can profitably be made with geometry of a more formal kind; and when the latter stage is reached, his practical knowledge should not only add life and interest to his theoretical work, but greatly accelerate its progress.

In each Section more exercises are provided than Teachers are likely to need for a first course: the rest may be taken afterwards with the corresponding propositions in the *School Geometry*, to which this little book is intended as a supplement as well as an introduction.

H. S. HALL.
F. H. STEVENS.

December, 1904.

CONTENTS.

NECESSARY INSTRUMENTS.

THE pupil should be provided with the following instruments and apparatus:

1. A flat ruler, one edge being graduated in centimetres and millimetres, and the other in inches and tenths.

2. Two set squares; one with angles of 45°, and the other with angles of 60° and 30°.

3. A pair of pencil compasses.

4. A pair of dividers, preferably with screw adjustment.

5. A semi-circular protractor.

6. Tracing paper. Squared paper.

It is also very desirable that pupils should have an opportunity of seeing and handling Models of the simpler Solid Figures.

A set of Models for use with this book has been specially prepared, and may be obtained (price 6s., in box, including carriage to any part of the United Kingdom) direct from the manufacturer

G. CUSSONS
The Technical Works
Lower Broughton
MANCHESTER

I. SOLIDS. SURFACES. LINES.

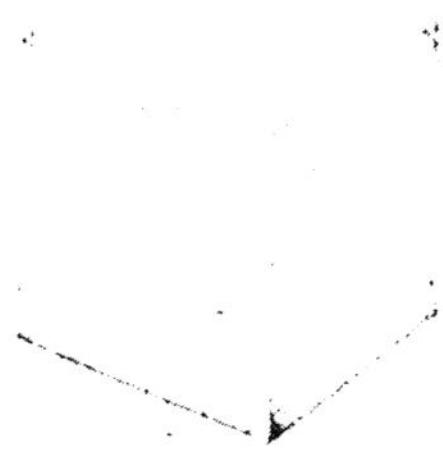

FIG. 1.

FIG. 2.

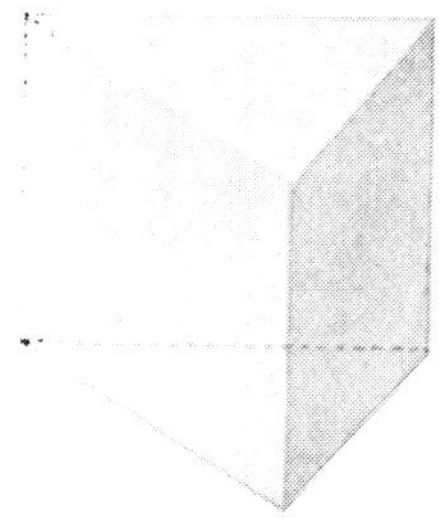

FIG. 3.

FIG. 4.

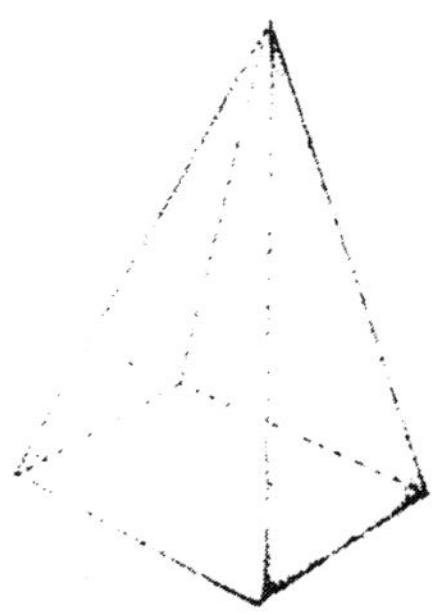

FIG. 5.

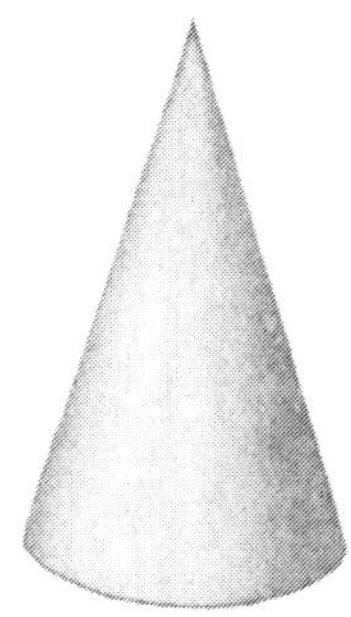

FIG. 6.

We have here some wooden models of what are called **solids** or **solid figures**, and they are differently named according to their shapes. That, for instance, of which a drawing is given in Fig. 1, is called a **cube**; that shewn in Fig. 2 is a **sphere**; that in Fig. 4 is a **cylinder**; and that in Fig. 5 is a **pyramid**.

The *outside* of these solid models, the part which we see and touch, is called the **surface**.

Sometimes the surface of a solid is all in one piece, as in the sphere (Fig. 2). Sometimes it consists of several parts: for instance in the cube (Fig. 1) the surface consists of *six* parts, all flat; these are called **faces**. Again, in the cylinder (Fig. 4) the surface consists of *three* parts, one rounded and the other two flat. Once more, the surface of the cone (Fig. 6) is in *two* parts, one rounded and running to a point, the other flat.

Let us now see how two neighbouring parts of a surface meet. They meet in **edges** or **lines**; and these lines are sometimes *straight*, and sometimes *curved*. In the prism and pyramid (Figs. 3 and 5) two neighbouring flat faces meet in a *straight* line; while in the cylinder (Fig. 4) the rounded part of the surface meets each flat end in a *curved* line.

How do the *edges* of a solid meet? If two edges meet at all, they meet at a **point**; as you will see if you look at the edges of a cube or pyramid (Figs. 1 and 5).

You now know what a solid is, and what a surface is; and you have learned that surfaces, or parts of a surface, meet in lines; and that lines meet in points. We have now to see how lines and points are represented in geometry; how *straight* lines are distinguished from *curved* lines; and how flat surfaces are distinguished from rounded ones.

Points. The smallest dot you can make on your paper with a sharp pencil, or with a fine needle, will give you an idea of what is meant by a geometrical point. A point is so minute that we do not think of its length, breadth, size, or shape: all we have to consider is its *position*.

As we have seen, a point marks the place where two lines cross one another. Points are named and distinguished from one another by attaching letters to them: thus we speak of the point A, or the point B.

•A

×B

Lines. We represent a line by drawing the point of a sharp pencil over a surface, such as a sheet of paper: this shews that *a line is traced out by a moving point.*

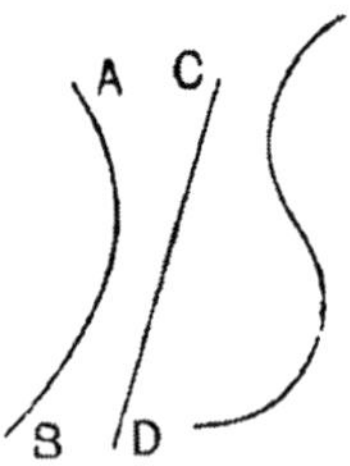

Several kinds of line are shewn in the margin.

All lines have *length*, some more, some less; but the *breadth* of a well drawn line is so small that no notice is taken of it in geometrical work: indeed, the finer your pencil-trace, the better it represents a line.

What we have to consider in a line is its *length* and *position*, and whether it is *straight* or *curved*. A line is named by two letters: thus we speak of the line AB, or the line CD.

Straight lines. No doubt you already know the meaning of the word *straight* well enough to give examples of straight lines. A very fine thread tightly stretched is a good instance of a straight line; so are the edges of the set squares which you are to use as rulers. But *straightness* needs some further illustration.

(i) When you walk along a winding lane you notice that your direction is continually changing; and if, for instance, you faced North when you started, you may presently find yourself facing East. But when you walk along a *straight* road, there is no change of direction as you advance; and if you faced North at starting, you will continue to face North.

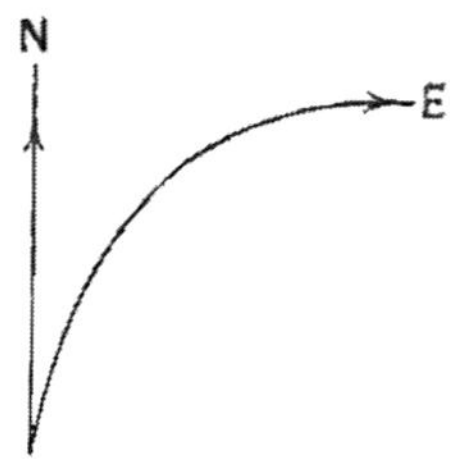

(ii) In a field there are two trees whose positions are marked by the letters A and B. Suppose you wish to go from one tree to the other by the *shortest* way. You can see at once what course you must steer. You must go *straight* from A to B. There are numberless *curved* lines along which you could go from one tree to the other, but the shortest way of all is the *straight* line. You notice that we have said *the* straight line; for you can see for yourself that there can only be *one straight* line leading from A to B.

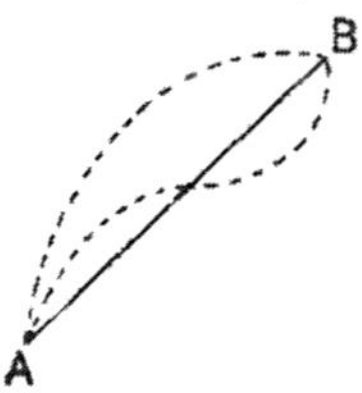

(iii) A strip of ground has been enclosed by two fences. One of these, AB, is straight: can the other be straight also? Clearly not; for we have already seen that there cannot be more than one *straight* line between A and B, though many curved lines such as ACB.

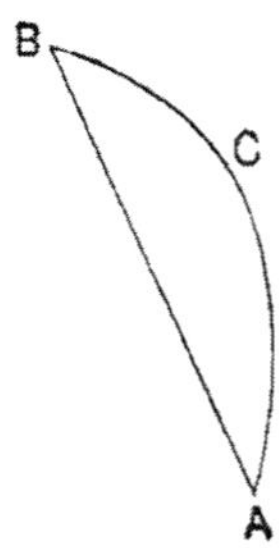

(iv) We will draw a curved line, and call it AB; then we will rule a straight line CD across it. You see that you can place your ruler so that the straight line will cut the curved one at *two* points, perhaps even more than two. Now take a *straight* line AB, and rule another straight line CD across it. Can you now place your ruler so as to cut AB in more than one point? You will soon find that you cannot.

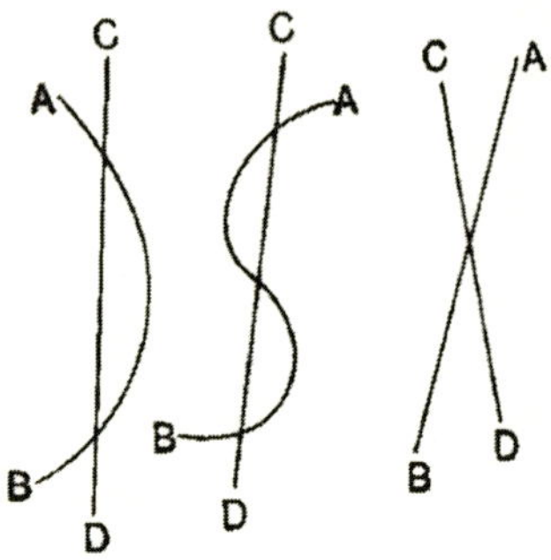

Let us now put together what we have learned about straight lines.

(i) *A straight line has the same direction throughout its length.*

(ii) *The straight line which joins two points is the shortest distance between them; and there is only one such straight line.*

(iii) *Two straight lines cannot enclose a space.*

(iv) *If two straight lines cross one another they can only cut at one point.*

When you rule a straight line between two points A and B, you are said to **join** AB.

Test of straightness. We can find if a given line AB is straight or not by means of a copy of it made on tracing-paper. If by turning the tracing either *round* or *over* we can in any way make the given line and the tracing enclose a space, then the given line is not straight. But if in *all* such positions the tracing can be made to fit exactly over the given line throughout its whole length, then we may conclude that the latter is straight. Apply this test to the two lines drawn below.

Planes. Several different kinds of surfaces have been shewn to you, and you have noticed that some are rounded or curved, and some are **plane**, that is to say, *flat*. How can we tell a plane surface from a curved one?

Lay the straight-edge of a ruler on a table, and notice that the *whole length* of the edge always rests upon the surface, *in whatever position the ruler is placed.* But if the ruler is placed in the hollow of a basin, only the ends rest on the surface: or again, if the straight-edge is laid against a sphere, it touches the surface at one point only.

Thus *a surface is* **plane** *when the straight line joining* **any** *two points on it lies entirely on the surface.*

NOTE. There are some curved surfaces, such as those of a *cylinder* and *cone*, along which a ruler will lie in *certain directions*, but not in *all* directions. The teacher should illustrate this with his models.

Ex. 1. What is the least number of *straight* lines that can enclose a space?

Rule *three* straight lines so as to enclose a space.

Rule *four* straight lines so as to enclose a space.

Ex. 2. Can two *curved* lines enclose a space? If so, make a drawing either free-hand or with compasses, shewing a space enclosed by two curved lines.

Ex. 3. Can *one* curved line enclose a space? Make a drawing to illustrate your answer, either free-hand or with your compasses.

Ex. 4. Mark a point on your paper, and call it A. How many straight lines, having different directions, can be drawn through the point A?

Rule *five* straight lines passing through A.

Ex. 5. Mark two points A and B. *Join* AB. Observe that the position of a *straight* line is fixed if we know *two* points through which it passes. How many *curved* lines can be drawn from A to B? Draw *three* such lines, either free-hand or with your compasses.

Ex. 6. Mark *three* points A, B, and C, placing them so that they do not lie all in a straight line. How many straight lines can be drawn by joining these points in pairs? Draw all these lines.

Ex. 7. Repeat Exercise 6, but take *four* points A, B, C, and D, no three of which lie in a straight line, and join them in pairs.

II. MEASUREMENT OF STRAIGHT LINES.

In practical geometry you will frequently have to measure the lengths of the lines you draw. For this purpose you have a scale which shews *inches* along one of its edges, each inch being divided into 10 equal parts: along another edge *centimetres* are marked, and each centimetre is also divided into 10 equal parts or *millimetres.*

Begin by carefully noticing the length of 1 inch and of 1 centimetre, so that you may be able to guess pretty nearly (even without measurement) how many inches or how many centimetres there are in a given line.

In writing down your measurements use the following abbreviations :

in. for *inch* ; *cm.* for *centimetre* ; *mm.* for *millimetre.*

Inches may also be denoted by the mark ("). Thus 3" means 3 *inches.*

The units on your scale are divided into *tenths* in order that your measurements may be recorded *decimally* : Thus

(i) *Three and seven-tenths inches* should be written 3·7 in., or 3·7".

(ii) *Eight-tenths of an inch* should be written 0·8 in., or 0·8".

(iii) *Five centimetres four millimetres* should be written 5·4 cm.

Ex. 1. Measure the lengths of **AB** and **CD** in inches and tenths of an inch.

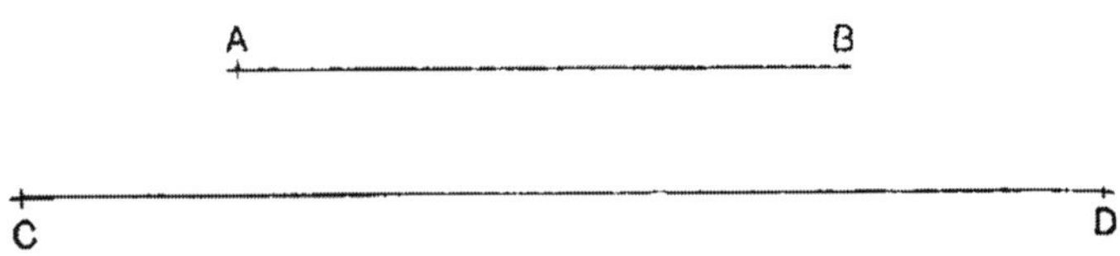

Ex. 2. Measure the above lines **AB** and **CD** as nearly as you can in centimetres and millimetres.

Ex. 3. Measure AX and XB in inches and tenths of an inch, and add your results together. Test your work by measuring AB.

Record your results thus :	By measurement, AX =	in.
	„ „ XB =	in.
	By addition, AX + XB =	in.
	By measurement, AB =	in.

Ex. 4. Measure AX and XB in centimetres and millimetres, and find their difference. Test your result by measuring AB.

Record your results as above.

Ex. 5. (i) Measure AB, AX, and XY in inches and tenths of an inch : hence reckon the length of YB, and test your result by measurement.

(ii) Measure AY, YB, and XB in centimetres, and hence find AY + YB − XB. What line should you now measure to test your result?

In each case arrange your results in tabular form.

Ex. 6. Draw straight lines to shew the following lengths :

2·6 in.,	5·0 cm.,	1·8″,	4·7 cm.,	0·8 in.
8·2 cm.,	3·1″,	0·7 cm.,	9 mm.,	33 mm.

(*Subdivision of a line by measurement.*)

Ex. 7. How would you find the middle point in the length of a strip of paper (i) by folding, (ii) by measurement?

Ex. 8. Draw a line AB of length 3″. What is the length of half AB? From AB mark off one-half, and thus find O the middle point of AB. Test your work by measuring OB.

A straight line is said to be **bisected** when it is divided into *two equal* parts.

Ex. 9. Draw a line AB of length 8·1 cm. What is the length of one-third of AB? With your dividers step off along AB one-third of its length, and thus divide AB into three equal parts.

A straight line is said to be **trisected** when it is divided into *three* equal parts.

Ex. 10. Draw a line AB of length 7·2 cm. By measurement, as explained above, cut off from it AP equal to *half* AB, and AQ equal to *one-third* AB. Find with your dividers how many times PQ is contained in AB. Explain your result by finding the value of $\frac{1}{2}-\frac{1}{3}$.

(*Comparison of* 1 *inch with* 1 *centimetre.*)

Ex. 11. Take 1 inch in your dividers, and apply them to your centimetre scale. How many centimetres and millimetres do you find in 1 inch?

It is impossible even with the greatest care to measure a length with perfect correctness; but the error is likely to be smaller *in proportion* in measuring a longer than in measuring a shorter length.

Ex. 12. Find the length of 1 inch in centimetres by measuring a length of 4 inches, and then dividing the result by 4.

Thus 4 inches = cm.

∴ 1 inch = cm.

Ex. 13. Measure a length of 1 centimetre against your inch scale. Then measure a length of 10 centimetres, and divide the result by 10. Compare the two equivalents of 1 cm., and observe that the second is likely to be the more correct.

(*Distances represented by Lines drawn to Scale.*)

A map or plan is a small but exact flat copy of the country or ground it represents. Therefore by measuring on a map the distance between two dots which mark certain towns, we may reckon the real distance between the towns themselves, provided we know the *scale* on which the map is drawn. For instance, if 1 inch measured on the map stands for 10 miles, then 2″ stands for 20 miles; 4·5″ for 45 miles; and so on. Such a map is said to be drawn on *the scale of* 10 *miles to* 1 *inch.*

Ex. 14. The plan of an estate is drawn on the scale of 75 yards to 1 inch :

(i) What distance on the ground is represented by 3·6″ on the map?

Here 1 inch represents 75 yards ;
∴ 3·6 inches .... 75 yards × 3·6
=270 yards.

(ii) What length on the map will represent 405 yards?

Here 75 yards are represented by 1 inch ;
∴ 405 yards .... 1 inch × $\frac{405}{75}$
=5·4″.

Ex. 15. A plan is drawn on the scale of 100 metres to 1 centimetre :

(i) What actual distances are represented on the map by 4·0 cm., by 5·6 cm., by 0·8 cm.?

(ii) Draw lines to represent 450 metres, 720 metres, 580 metres, and 60 metres.

Ex. 16. On a map in which 1″ stands for 20 miles, the distance between Halifax and Hull is represented by 3·2″: What is the actual distance?

Bedford is 86 miles from Norwich: how far apart would they be on the map?

Ex. 17. The points marked **S*a*.**, **S*o*.**, **W** represent the positions of Salisbury, Southampton, and Winchester on a map whose scale is 10 miles to 1 inch.

Find by measurement and reckoning the actual distances between Salisbury and Winchester, Winchester and Southampton, Southampton and Salisbury.

[*In the following Exercises plans are to be drawn on squared paper ruled to tenths of an inch, and the results are to be got by measurement and reckoning.*]

Ex. 18. I walk 4 miles due North, then 3 miles due East. Draw a plan to shew my journey, making 1 in. stand for 1 mile; then by measurement find how far I am from my starting point.

Ex. 19. Draw the ground-plan of a room, 30 feet long by 20 feet wide, making 1″ represent 10 feet. Find as nearly as you can the actual distance between two opposite corners.

Ex. 20. An upright pole, standing 25 feet high, is stayed by a rope carried from the top to a point on the ground 15 feet from the foot of the pole. Represent this by a drawing (scale 10 feet to 1 inch); and find the length of the rope.

Ex. 21. A ladder reaches a window-sill 15 feet high, and the foot of the ladder rests on the ground 8 feet from the front of the house. Draw a plan (scale 5 feet to 1 inch), and use it to find the length of the ladder.

Ex. 22. Looking Eastward from my house, I see a church tower which I know to be 2 miles distant. Looking North I see a second tower $1\frac{1}{2}$ miles away. Draw a plan (scale 1 mile to 1 inch), and find how far the towers are apart.

Ex. 23. A ship on leaving harbour sails 22 miles South, then again 22 miles West. Represent her course on the scale of 10 miles to 1 inch, and find her distance from the harbour.

Ex. 24. In rowing across a river 48 metres wide, a man was carried 16 metres down stream. Represent this on a plan (scale 20 metres to 1 inch); hence find the distance between the starting-point and landing-point.

III. STRAIGHT LINES CONTINUED.

⁂ *This Section may be postponed for revision.*

If you measure the same line in several ways, some of your results may be a little too large and some a little too small. The *average* of your results is likely to be nearer the truth than any single result. To find the average, add your results together, and divide their sum by the number of them.

Ex. 1. Measure AB in inches and also in centimetres; and hence express 1 inch in terms of cm. and mm.

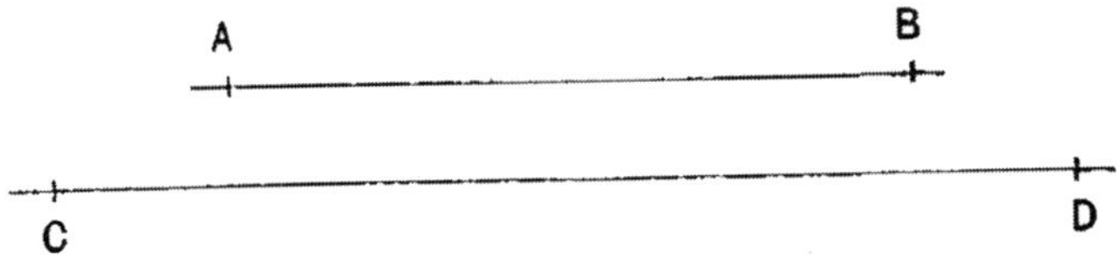

Measure CD, and repeat the process. Now find the average of your two results.

(Judging Lengths. Errors. Relative Errors.)

It is important that you should train your eye to subdivide any unit of length into *tenths* without actual measurement. Remember that *one-half* = *five*-tenths: this gives a standard to judge by. Fix your eye on the middle point, and mentally divide each half into five equal parts.

Ex. 2. The lines marked AB are all 1 inch long. State in each case how many tenths of an inch there are in AP; then verify your answer by measurement.

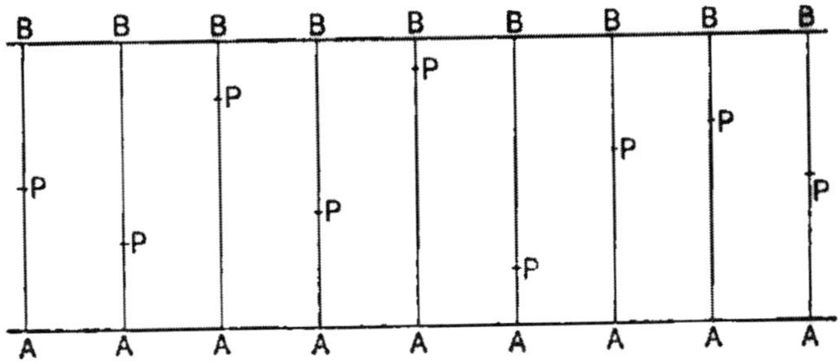

Ex. 3. Draw six lines each 1 inch long, calling one end A. Then mark a point P in each (without measurement) so that, as nearly as you can judge, AP may be in succession 0·4″, 0·7″, 0·2″, 0·9″, 0·3″, 0·6″.

Check your attempts by measurement.

Ex. 4. (i) Judge as nearly as you can in inches and centimetres the lengths of the lines given below.

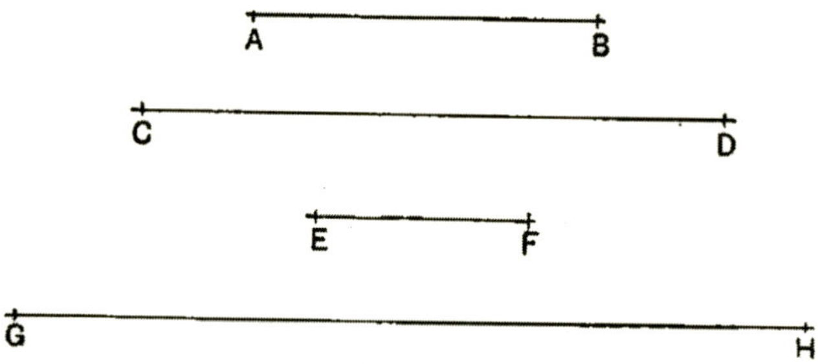

Check your estimates by measurement, and tabulate the results as below, leaving the last column blank for the present.

	Measured length.	Estimated length.	Actual error.	Percentage error.
AB {	in.	in.	in.	
	cm.	cm.	cm.	

*** *Other lines of greater length and not all horizontal should be given by the teacher.*

(ii) Draw lines as nearly as you can judge without measuring to shew 6 cm., 2·0″, 8 cm., 3·5″. Measure your attempts; note your errors, and tabulate the results.

In judging the importance of an error we do not care so much whether it is large or small, as whether it amounts to a large or small fraction of the quantity we are estimating. For instance: suppose that in guessing the length of a line whose real length is 5 cm. we are wrong by 1 cm.; while in guessing a line 20 cm. long we are wrong by 2 cm. The actual error in the latter case is greater than in the former, but it is really of less importance. For in the second case the error is only *one-tenth* of the real length, that is, *one in ten*; while in the first case it is *one-fifth*, or *one in five*. Errors thus measured as fractions of the true value are called **relative errors**: and it is convenient to reduce them to a fixed standard, as so many *in one-hundred*, or so many *per cent.* Take the following case:

Real length.	Estimated length.	Actual error.	Percentage error.
8·0 cm.	7·5 cm.	0·5 cm.	

Here on a real length of 8 cm. the error is 0·5 cm.

∴ .... 100 cm. .... 0·5 cm. × $\frac{100}{8}$

$=6\frac{1}{4}$ cm.

That is, the error is at the rate of $6\frac{1}{4}$ *in one hundred*, or $6\frac{1}{4}$ *per cent.* We may now enter $6\frac{1}{4}$ in the last column.

Ex. 5. Fill up the percentage column in Ex. 4, giving the percentage correct to one decimal figure.

Hitherto the lines which you have had to measure in inches and tenths of an inch have contained an exact number of tenths. This will not always be so. For example

the line AB is more than 2·4″ and less than 2·5″. In this case we may mentally divide the tenth in which B falls into *ten* equal parts, that is to say, into *hundredths of an inch*, and judge as nearly as we can how many of these hundredths are to be added to 2·4. In this instance about *seven*-hundredths should be added, so that the length of AB is nearly 2·47″.

Ex. 6. Draw on squared paper a figure like that in the margin, making OA and OB each 2″ long. Put P, Q, R and X, Y, Z at the half-inch divisions; then measure AB, RZ, QY, PX as nearly as you can in *inches*, *tenths* and *hundredths*.

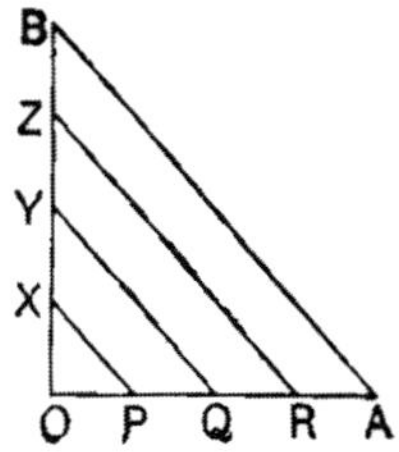

IV. CIRCLES.

Mark a point O on your paper. Take a distance of 5 cm. between the points of your compasses; then, placing the steel point at O, turn the compasses between your fore-finger and thumb so as to draw a curved line with the pencil-point.

As the curved line is being traced out, notice carefully that the pencil-point always keeps the same distance from O. What distance? Notice also that the pencil returns to its starting point, so as to close the curve. Why is this?

The curve you have thus drawn is called a **circle**, and the point O is its **centre.** Sometimes the word *circle* means the space enclosed by the curve, and then the curve itself is said to be the **circumference** of the circle.

Ex. 1. Mark a few points, say four, anywhere on the circumference of the circle you have drawn: call them A, B, C, D. Join OA, OB, OC, OD. How do you know that these lines are all equal? Tell their length without measuring them.

Straight lines drawn from the centre of a circle to its circumference are called **radii.** All the radii of a circle are equal.

Ex. 2. Mark a fixed point O on your paper: then with your compasses mark any *four* points whose distance from O is 2·0″. How many points could you mark whose distance from O is 2·0″? Draw a curve to pass through all of them.

Ex. 3. Suppose a point X is taken 1·7″ from the centre of the circle you have just drawn (Ex. 2); another point Y is 2·0″, and a third point Z is 2·3″ from the centre. Which of these points is on the circumference? Which outside it? Which within it?

Ex. 4. Invent some other means, besides compasses, by which a circle could be drawn having a fixed point O as centre.

Ex. 5. Now explain in your own words what a circle is, telling how the circumference is related to the centre.

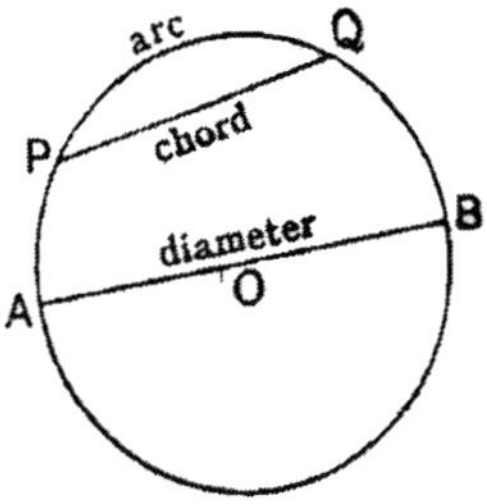

Taking a point O as centre, draw a circle with a radius of 1·5″. Then through the centre O draw any straight line ended each way by the circumference. Such a line is called a **diameter,** and is represented in the Figure by AB.

Ex. 6. What is the length of AB in your drawing? Answer this without measuring? Are all diameters of a circle equal?

Now carefully cut your circle out, and fold it about the diameter AB, thus dividing the circle into two parts. Do you find that one part fits exactly over the other? If so, this shews that the two parts *are of the same size and shape.* Flatten out the circle; rule any other diameter, and fold the circle about it as before. Again you find that one part fits exactly over the other. All this we express by saying that a circle is **symmetrical** about any diameter.

The two equal parts into which a circle is divided by a diameter are called **semi-circles.**

An **arc** (i.e. *bow*) is any part of the circumference of a circle

A **chord** (i.e. *string*) is the straight line joining the ends of an arc.

Ex. 7. Draw a circle of diameter 3·0″, and on the circumference mark a point X. From X draw two chords, one 1·5″ long, the other 2·0″ long. What is the length of the longest chord in this circle?

Ex. 8. In the above Figure notice that the chord PQ divides the circumference into *two* arcs. Point them out. Can a chord ever cut off two *equal* arcs? Which is the longer line, an arc, or the chord which joins its ends?

(*Two or more circles. Intersection of circles.*)

Ex. 9. Mark a point O on your paper, and from O as centre draw three circles, one of radius 3·5 cm., the next of radius 4·0 cm., the third of radius 4·5 cm. Notice that the circumferences do not cross or cut one another. Why not?

Circles which have the same centre are said to be **concentric.**

Ex. 10. (i) Take two points A and B, 7 cm. apart. With A as centre draw a circle of radius 4 cm.; and with B as centre draw a circle of radius 2 cm. Explain why each circle is outside the other. What is the shortest distance between the circumferences?

(ii) Again take two points A and B, 7 cm. apart; and, as before, with A as centre draw a circle of radius 4 cm. But this time draw from centre B a circle of radius 5 cm. Why do these circles overlap? At how many points do the circumferences cut one another?

(iii) Once more take two points A and B, 7 cm. apart, and with A and B as centres draw two circles, one of radius 4 cm., the other of radius 3 cm. Do the circumferences cross one another? Do they meet? If your work is carefully done, the two circles just *touch* one another. Where is the touching point? Say why.

Ex. 11. Take two points A and B, 2 cm. apart; and with centre A draw a circle of radius 5 cm. With centre B draw a circle of radius 3 cm. How does this circle meet the first, and where is the meeting-point?

Ex. 12. Can you draw two circles which cut one another at more than two points? Try.

Ex. 13. Take two points 3″ apart, and call them A and B. With centre A and radius $2\frac{1}{2}$″ draw a circle. With centre B and radius 2″ draw a second circle. Call the points at which the circles cut one another P and Q. How far is P from A and from B? How far is Q from A and from B?

Ex. 14. Take two points A and B, 8 cm. apart. Find with your compasses a point which is 6 cm. from A and also 6 cm. from B. Can you find more than one such point? How many?

Ex. 15. Draw a line 2·5″ long, and find with your compasses a point that is 2·0″ from each end. How many such points are there?

Ex. 16. Take two points X and Y, 9 cm. apart. Find a point which is 6 cm. from X and 5 cm. from Y. How many such points are there?

Ex. 17. Draw a line 3·3″ long, and find two points each of which is 2·2″ from one end and 1·8″ from the other.

Ex. 18. Two forts defending the mouth of a river, one on each side, are 10 kilometres apart: their guns have an effective range of 6000 metres. Draw a plan (scale 1 km. to 1 cm.) shewing what part of the river is exposed to fire from both forts.

PROBLEM 1.

To bisect a straight line AB *with ruler and compasses.*

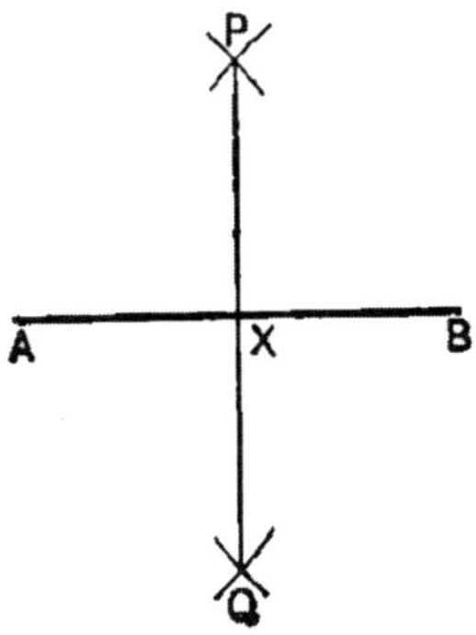

[The given straight line AB may be of any length: about 3″ to 4″ will be convenient, but do not measure it.]

Construction. Take in your compasses any length that appears to you to be greater than half AB (say about $2\frac{1}{2}''$); and then with centre A draw arcs on each side of AB.

Again with centre B, and with the *same radius* as before, draw arcs to cut the first arcs as shewn in the Figure. Call the cutting points P and Q.

Join PQ, and put X at the point where this line crosses AB.

Now take AX in your dividers, and see if BX is equal to it.

(*Further Tests.*)

(i) Mark the points A, B, and X on tracing-paper, and turning it round, place the trace of A on B, and the trace of B on A. Where does the trace of X fall? How does this experiment shew that AB has been bisected at X?

(ii) If the arcs drawn from centre B had a greater radius than those drawn from centre A, would X still be the middle point of AB? If not, towards which end of AB would X lie? Take your compasses and try. You see then that X is the middle point of AB *because we have worked from centre* B *in exactly the same way as from centre* A.

(iii) Why did we take a radius *greater* than half AB? What would have happened if the radius had been *less* than half AB? or exactly half AB? Take your compasses and try.

Ex. 19. Draw a line 8·5 cm. long, and bisect it with ruler and compasses. Test your drawing with the dividers.

Ex. 20. Draw a line 3·4″ long. Find the middle point X *by measurement*. Now bisect AB *by construction*, and see if the line PQ passes through X.

Ex. 21. Draw a line 9·6 cm. long. Bisect it by construction; then bisect each half.

Draw a circle, say of radius 2·0″, and with *the same radius* mark off points round the circumference. How many steps can you thus take? **Six** exactly. Are the *arcs* which you thus cut off each 2″ in length? Are they more or less than 2″? Join the points of division in order. Are the *chords* each 2″ in length? Why so? Join the centre to each point of division, and thus complete the pattern shewn in the margin.

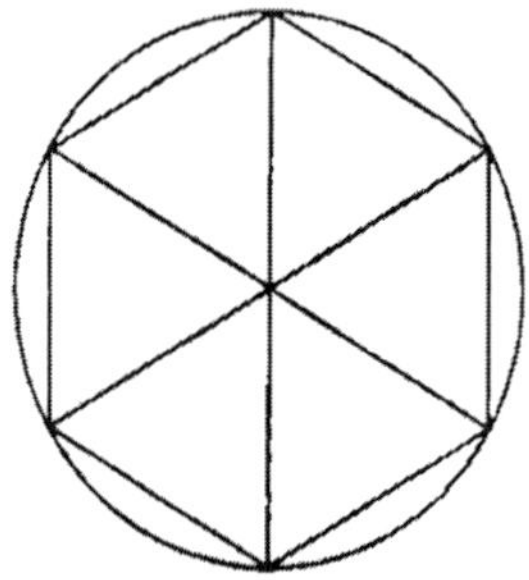

Ex. 22. Invent some simple experiment, for instance by cutting out, or folding, or by means of a tracing, to shew that the six arcs are of equal length (though not 2″). Try to find the length of one of these arcs by laying a thread along it, straightening the thread out before measurement.

Ex. 23. Draw the patterns of which small copies are given below. Your drawings should be twice the size of the copies.

Fig. 1.

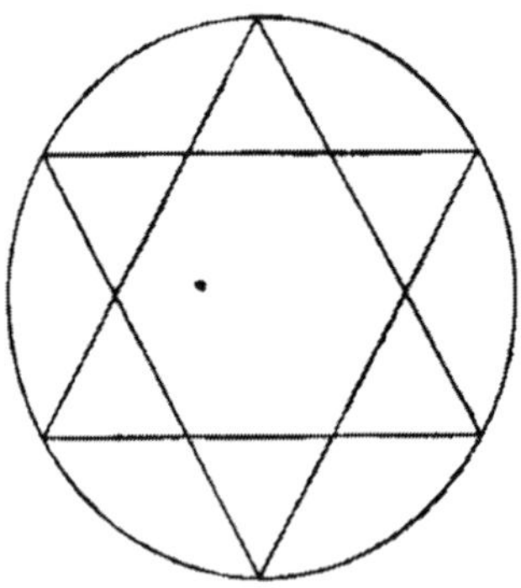

Fig. 2.

V. ANGLES.

Any two straight lines drawn from a point O form what is called an **Angle.**

The point O is the **vertex,** and the lines are the **arms** of the angle.

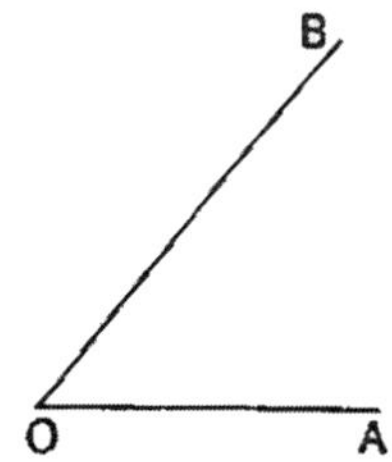

Put A at any point on one arm, and B at any point on the other; then the angle at O is named either by the letters AOB or by BOA, the letter O at the vertex being between the other two.

The sign "∠" is used for the word *angle.*

Thus the angle in the Figure is called the ∠AOB or the ∠BOA.

Ex. 1. Draw two straight lines forming an angle at the point O. Put A and P at any two points in one arm, and B and Q at any two points in the other arm. Then name the angle at O by three letters in all the ways you can.

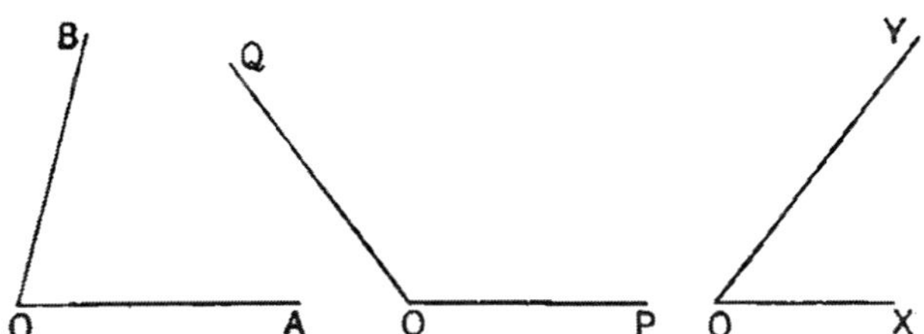

Fig. 1. Fig. 2. Fig. 3.

Figs. 1, 2, and 3 represent three angles. In Fig. 2 you see that the arms are *more widely opened out* than in Fig. 1; while in Fig. 3 the arms are *less widely opened out* than in Fig. 1. This we express by saying that

the angle POQ is *greater than* the angle AOB;

the angle XOY is *less than* the angle AOB.

Thus the size of an angle does not depend on the length of its arms, but only on the *slope* or *inclination* of one arm to the other.

How can we find out whether the angle ABC *is equal to* the angle XYZ? Here is one way.

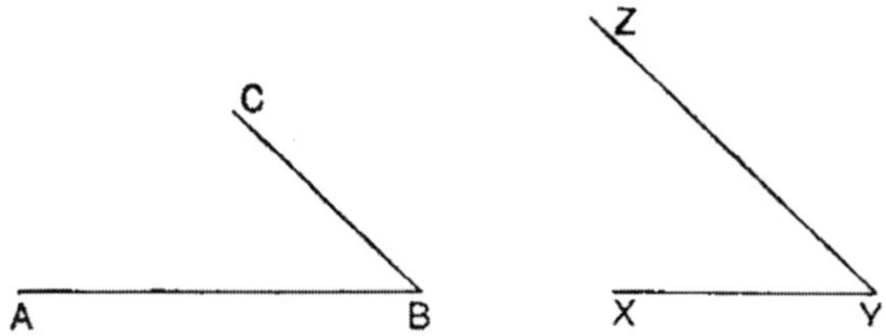

Copy the angle ABC on tracing-paper. Move the tracing so that the vertex B comes over the vertex Y; then place the trace of BA *along* YX. This you can always do whether the two angles are equal or not. Now observe where the trace of BC falls. Does it lie *along* YZ? If so, the angles ABC, XYZ are equal, though their arms are not of the same length.

What conclusion would you have drawn if BC had fallen *within* the angle XYZ? Or again, if BC had fallen outside the angle XYZ?

Ex. 2. Draw two angles making them equal to one another as nearly as you can judge, but do not make the arms of the same length. Try with tracing-paper if the two angles are really equal; and if not, say which is the greater.

Ex. 3. Draw two angles, one greater than the other. Give the larger angle shorter arms than the smaller one.

Take your compasses, and holding one leg fixed along the desk, open them gradually out. Observe that you make the other leg *rotate about the pivot* like the hand of a watch, and that as you do so, you constantly increase the angle between the legs.

We may thus suppose an angle AOB to be formed by a *fixed* line OA and a *rotating* line OB, the size of the angle AOB being given by the *amount of turning* required to bring the rotating arm from its first position OA to its subsequent position OB.

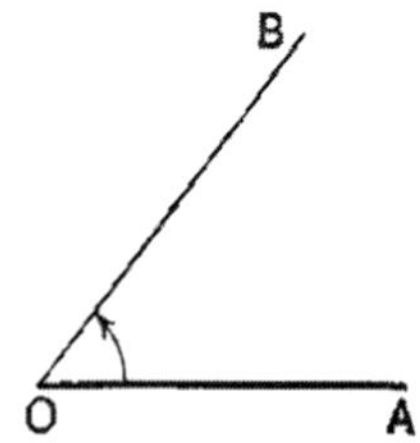

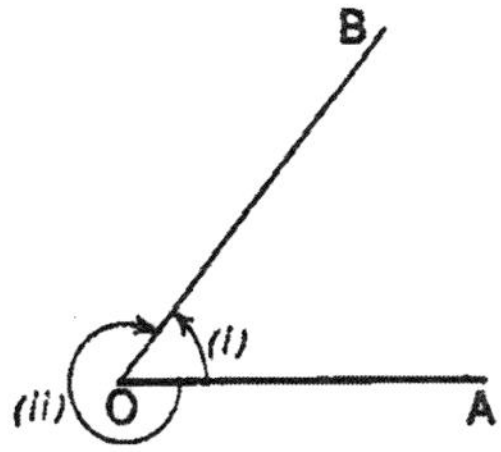

Ex. 4. When two straight lines OA, OB meet at a point O, *two* angles are formed. The first is got by supposing OB to have moved from OA into its present position by turning the *shorter way round*, marked (i); the other by supposing OB to have turned the *longer way round*, marked (ii). The latter angle is said to be **reflex**. Illustrate this by drawing any angle; then place one end of a penholder at the vertex, and turn it from one arm to the other in opposite ways. Which way gives the reflex angle?

Unless the word *reflex* is specially used, *the angle at* O will always mean the smaller of the two angles formed by the arms.

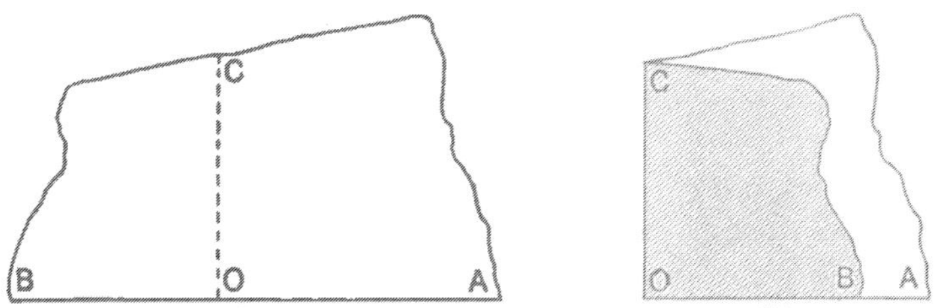

Fig. 1. Fig. 2.

Take a piece of paper having a straight edge AB (Fig. 1). Fold, as in Fig. 2, so as to bring the point B over on to the straight edge towards A. Open out the paper, and mark the crease OC. The angles AOC, BOC are equal. Why so?

Try the experiment two or three times, and fit together the folded papers. Do you find that all the angles you get in this way are of the same size?

In each case you have a straight line OC meeting a straight line AB in such a way that the angles made on either side of OC are equal. Such angles are called **right-angles**; and our experiments shew that all right angles are equal. Thus a right angle may be taken as a standard with which to compare other angles.

OC is said to be **at right angles to** AB; or **perpendicular** to AB.

A right angle is divided into 90 equal parts called degrees (°).

That is, *one right angle* $= 90°$.

An **acute** angle is less than one right angle.

An **obtuse** angle is greater than one right angle.

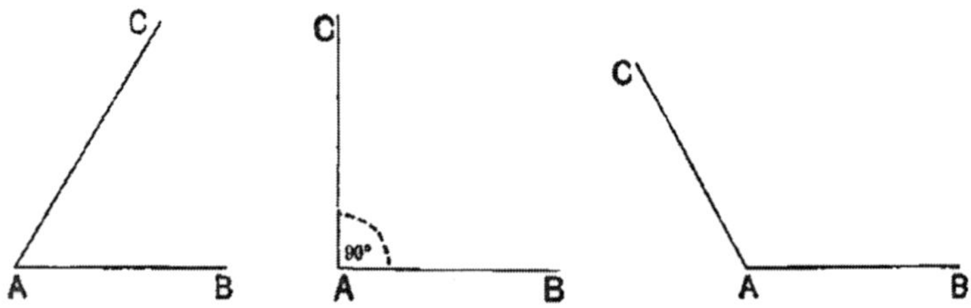

acute angle right angle obtuse angle

Ex. 5. Fold a piece of paper of any shape, and call the straight folded edge AB. Then (without opening the paper out) fold again so as to bring B over A. On unfolding, the creases cross one another, forming four angles. What can you tell of these angles? Are they equal? Are they right angles? Say why.

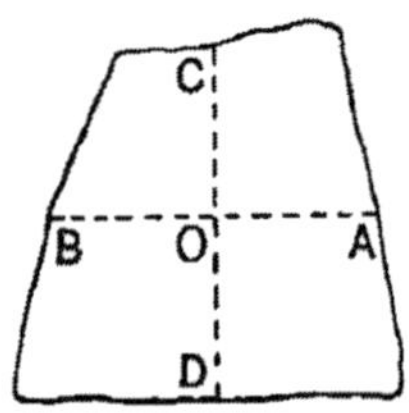

Ex. 6. A line, starting from the position OA, rotates about O; and having made *a complete revolution*, returns to OA. Through how many degrees has it revolved?

Through how many degrees does the line revolve in making *one quarter* of a revolution? In making *one half* a revolution?

Observe that a **complete** revolution corresponds to 4 right angles.

a **quarter** revolution .... 1 right angle.

a **half** revolution .... 2 right angles.

Ex. 7. Through how many degrees does the minute-hand of a clock revolve in $\frac{1}{4}$ hour, in $\frac{1}{2}$ hour, in $\frac{3}{4}$ hour, in 1 hour?

Ex. 8. Through how many degrees does the minute-hand revolve in 5 minutes, in 25 minutes, in 36 minutes? How long will it take to turn through 48°? Through 102°? Through 9°?

Ex. 9. If a wheel makes 10 revolutions a minute, through how many degrees will it turn in 1 second?

(Angles at the Centre of a Circle.)

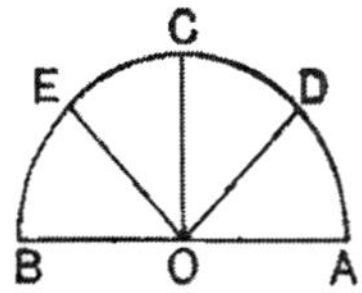

Ex. 10. In the marginal figure the angles AOD, DOC, COE, EOB have been made all equal. How many degrees are there in the ∠AOD? In the ∠AOC? In the ∠AOE? In the ∠AOB?

The angle AOB is called a **straight** angle.

Draw a semi-circle, radius 2″; cut it out, and obtain the lines OC, OD, OE by folding.

The figure formed by the radii OA, OC and the arc AC is called a **quadrant**, or quarter, of a circle. Point out another quadrant.

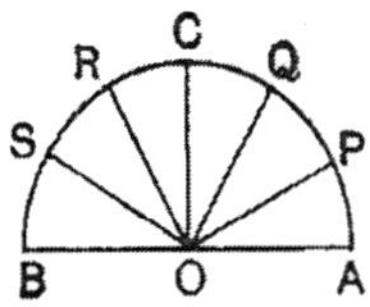

Ex. 11. In the marginal figure the angles AOP, POQ, etc., have been made all equal. How many degrees are there in each of the ∠ˢ AOP, AOQ, AOC, AOR, AOS, AOB? How many degrees in the ∠ˢ SOC, ROP?

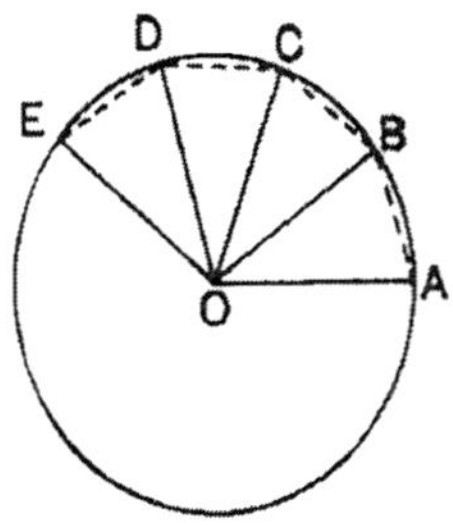

Ex. 12. Draw a circle of radius 5 cm. Take *any* distance you like in your compasses, and with this distance mark off points round the circumference. Call the points A, B, C, etc., and join them to the centre O. Now, what are the equal lengths you have been stepping off? Certainly equal *chords*, though you have not actually drawn them.

Try to invent some practical way (either by means of tracing, or by cutting out and fitting one part over the other) of finding if in measuring off equal *chords* you have also cut off equal *arcs*. Are the *angles* AOB, BOC, COD, etc. equal too?

Ex. 13. In the Figure of the last Exercise, how many times does the arc EA contain the arc BA? How many times does the ∠EOA contain the ∠BOA?

Thus in a circle (or in *equal* circles) you have found by experiment that when you measure off equal *chords*, you thereby cut off equal *arcs*; and by cutting off equal arcs, you can make equal *angles* at the centre. This principle is most important, and is used again and again in practical geometry.

(Use of the Protractor.)

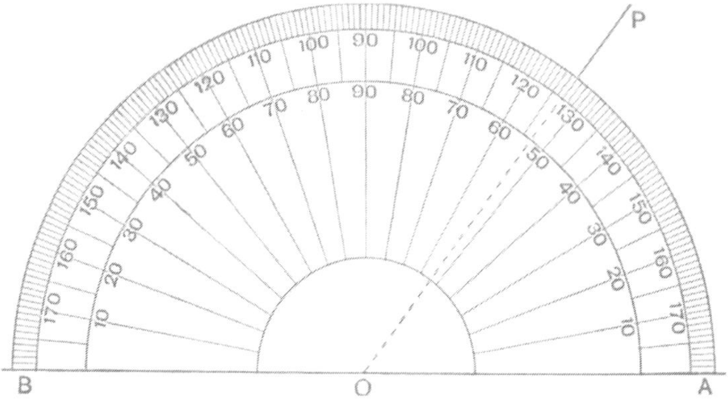

Your protractor shews a semi-circular arc divided into 180 equal parts, which for convenience are numbered from each end.

(i) *To measure the number of degrees in a given angle,* place the protractor with its centre at the vertex, and the diameter in line with one of the arms of the angle; then observe the mark of division on the rim under which the other arm passes.

(ii) *To make an angle of a given number of degrees (say 53°),* draw one arm **OA**; place the protractor with its centre on **O** and its diameter in line with **OA**; mark a point on your paper as close as you can to the 53rd division on the rim; remove the protractor and join the vertex **O** to the point so marked.

Ex. 14. Measure in degrees the angles **AOB** and **BOC**. Add your results together, and test by measuring the angle **AOC**.

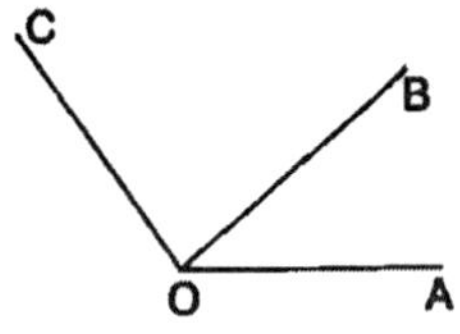

*** *Angles drawn with arms of sufficient length for use with the protractor should be given for measurement by the teacher.*

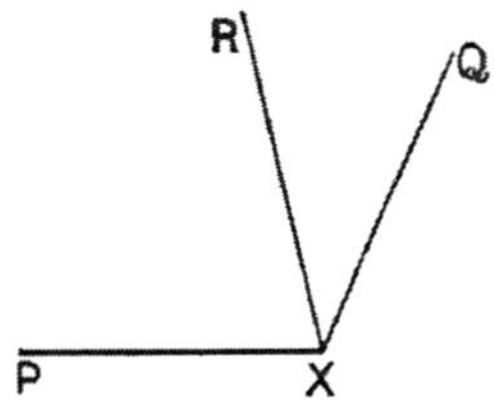

Ex. 15. Measure the angles PXQ and RXQ. Find by subtraction the number of degrees in the angle PXR. Test your result by measuring that angle.

*** *Other diagrams for practice in measuring angles should be provided by the teacher.*

Ex. 16. Draw a straight line AB of length 3″. From A draw a line making an angle of 62° with AB.

From B draw a line making an angle of 62° with BA. (Both lines are to be drawn on the same side of AB.)

Ex. 17. Repeat Exercise 16, but make the angles at A and B (i) 27°, (ii) 81°, (iii) 157°. (This may be done in a single figure.)

Ex. 18. Draw a straight line AB of length 8 cm. From A draw two lines, one on each side of AB, each making an angle of 47° with it. Repeat the process, making angles of 75° and 131° on each side of AB. (This is to be done in a single figure.)

If your figure were folded about AB, how would the lines on one side of AB fall with regard to those on the other side?

Ex. 19. Draw (i) an acute angle, (ii) an obtuse angle, (iii) a reflex angle.

In each case judge as nearly as you can (without using your protractor) how many degrees there are in the angle.

Check your estimates by measurement, noting your errors; and express these errors as percentages of the measured values.

Tabulate your results as in Ex. 4, p. 12.

Ex. 20. Without using your protractor draw angles as nearly as you can judge to contain 45°, 30°, 78°, 125°, 64°, 115°, 225°.

Measure your attempts, and tabulate the results.

(Adjacent and Vertically Opposite Angles.)

Two angles which have one arm in common, and lie on opposite sides of it, are said to be **adjacent**. Point out four pairs of adjacent angles in the marginal Figure.

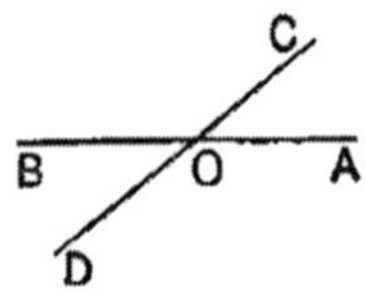

The angles AOC, BOD are said to be **vertically opposite**. Point out another pair of vertically opposite angles.

Ex. 21. Draw a straight line AB, and from any point O in it draw another line OC. Do this three times, placing OC in different positions.

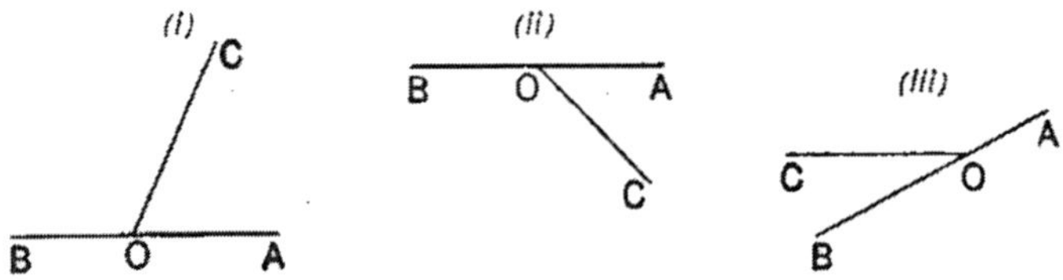

Measure the angle AOC; and, without moving the protractor, measure the adjacent angle BOC. In each case fill up the form:

$\angle AOC + \angle BOC =$ degrees $=$ right angles.

Compare the three results, and write down in words the conclusion you draw. Try to explain the reason.

Ex. 22. In the Figs. of Ex. 21,

(i) if the $\angle AOC = 65°$, *reckon* the $\angle BOC$.

(ii) if the $\angle BOC = 140°$, reckon the $\angle AOC$.

(iii) if the $\angle AOC = 153°$, reckon the $\angle BOC$.

Ex. 23. Draw a straight line OX. Make the angle XOP$=55°$; and on the other side of OX make the angle XOQ$=120°$. Are OP and OQ in one straight line? If not, how should OQ be turned, so as to bring it into line with OP?

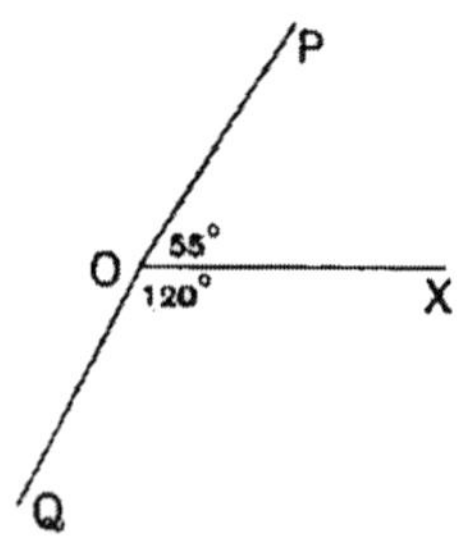

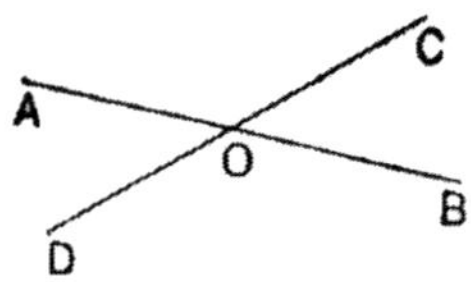

Ex. 24. Draw the straight lines AB, CD crossing one another at O. Measure the angle AOC. Hence reckon the angles BOC, AOD, DOB.

Now compare the vertically opposite angles thus:

$\angle$ BOC = degrees } $\angle$ AOC = degrees }
$\angle$ AOD = degrees } $\angle$ BOD = degrees }

Write down your conclusion in words.

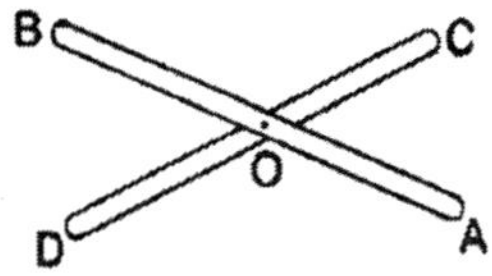

NOTE. The equality of vertically opposite angles should be illustrated by experiment.

For instance: two narrow strips of cardboard may be pivoted by a drawing-pin at O. Bring the strips into coincidence, then slowly open them out. Observe that the same movement which opens the angle AOC, also opens the angle BOD: that is to say, these angles are the result of the *same amount of turning*, and are therefore equal to one another.

Ex. 25. In the Figure of Ex. 24:

(i) If the $\angle$ BOD = 143°, reckon each of the $\angle$s BOC, COA, AOD.

(ii) If the $\angle$ AOD = 29°, reckon each of the $\angle$s DOB, BOC, COA.

(iii) If the $\angle$ COA = 137°, reckon each of the $\angle$s BOD, DOA, COB.

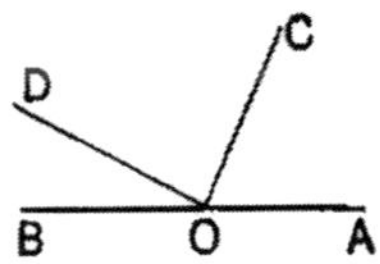

Ex. 26. Draw a straight line AB, and from a point O in it draw any straight lines OC, OD, on the same side of AB. Measure the angles AOC, COD, DOB, and find their *sum*. Account for the result.

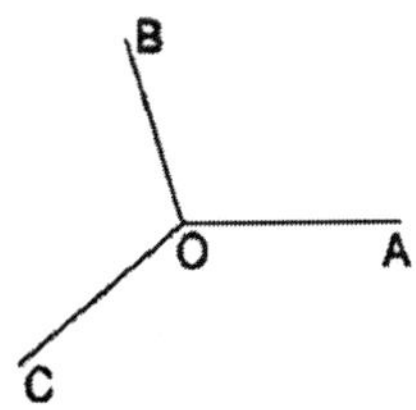

Ex. 27. From a point O draw three straight lines OA, OB, OC. Measure the $\angle$ AOB, BOC, COA, and fill up the following:

$\angle$ AOB + $\angle$ BOC + $\angle$ COA = degrees.
= right angles.

Ex. 28. In the Figure of Ex. 27:

(i) If $\angle$ AOB = 125°, and $\angle$ BOC = 82°, reckon the $\angle$ COA.

(ii) If $\angle$ AOB = 134°, and $\angle$ AOC = 152°, reckon the $\angle$ BOC.

In each case test by measurement.

VI. ANGLES CONTINUED.

CONSTRUCTIONS WITH RULER AND COMPASSES.

Problem 2.

To draw an angle equal to a given angle O.

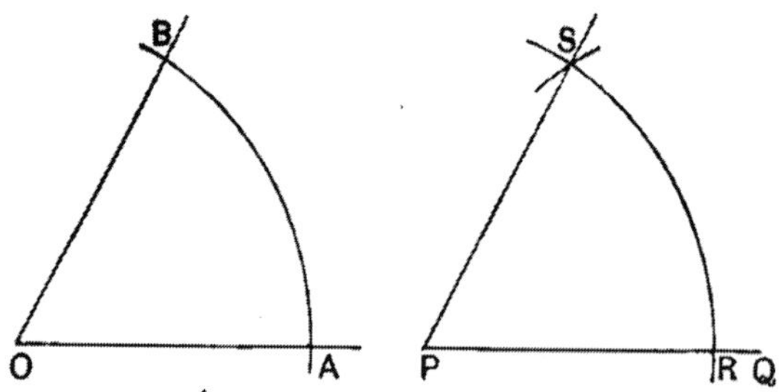

[The angle O may be of any size: its arms may be conveniently made about 8 cm. in length.]

Construction. Draw a straight line PQ, say about 8 cm. long.

With centre O, and any length (say 6 cm.) as radius, draw a circle cutting the arms of the given angle at A and B.

With centre P, and with the *same* radius as before, draw a circle cutting PQ at R.

(Only arcs of these two circles are shewn in the Figure.)

Take in your compasses the distance between the points A and B, that is to say, the length of the *chord* AB (there is no need to draw the chord): with centre R, and this length as radius, cut the second circle at S.

From P draw a straight line through S.

Now measure both angles with your protractor, and see if they are equal.

(*Further Tests.*)

(i) Trace the ∠QPS, and see if the tracing can be made to *coincide with*, that is, *exactly fit over* the given ∠O.

(ii) Now, having found by experiment that the two angles are equal, let us see *why* they are equal. In the equal circles, whose centres are at O and P, you measured off equal *chords*, though you did not draw them. Are the *arcs* RS, AB equal? How do you know this? And we have found by experiment (p. 23) that in equal circles, by joining the ends of equal arcs to the centres, we make equal angles.

(iii) Would the ∠RPS have come out equal to the ∠O, if you had drawn the two circles with *different radii*. Try: make the circle with centre O larger than the circle with centre P, but otherwise work as before. Are the angles at P and O equal now? Then which is greater?

Ex. 1. Draw an angle of 73° with your protractor. Then, with ruler and compasses only, construct an equal angle. Test your drawing with the protractor.

Ex. 2. Repeat the last Exercise with an angle of 126°.

Ex. 3. Draw an angle AOB of any size. Then, with ruler and compasses, draw a line OC making the ∠AOC equal to the ∠AOB on the other side of OA. Test with tracing-paper.

Ex. 4. Draw an angle of 35° with your protractor; then, with ruler and compasses, construct another angle *three times* the size of the first. Test your construction by measurement.

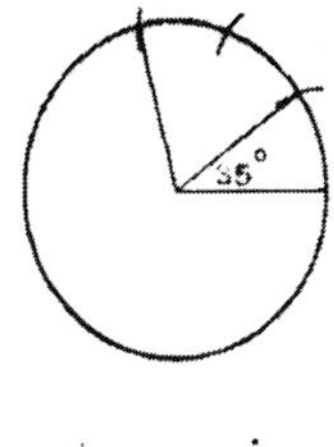

Ex. 5. I want to draw an angle *five times* as great as a given angle A. Explain in your own words how this may be done.

Ex. 6. Draw a circle with centre O and *any* radius. Step off this radius from A to B on the circumference, and join OA, OB.

What fraction is the ∠AOB of *four right angles*, and why? How many degrees are there in the ∠AOB? Answer, then test by measurement.

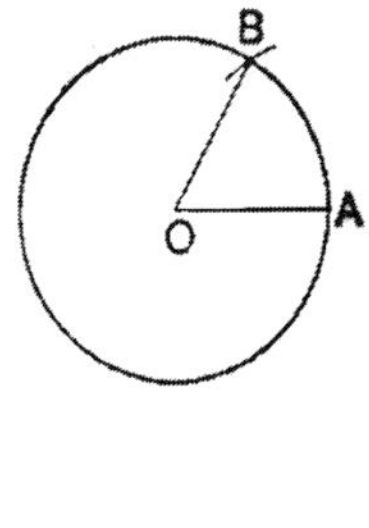

Ex. 7. Draw an angle of 120°, using ruler and compasses only.

Ex. 8. With your protractor draw a *right angle* AOB. With centre O and any radius (say 7 cm.) draw the arc AB. What part of the whole circumference is this arc?

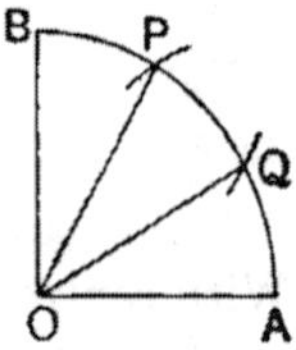

From centre A, with *the same radius*, cut the arc at P; and from centre B, with the same radius cut the arc at Q. Join OP, OQ.

How large are the ∠ˢ AOQ, QOP, POB? Answer, giving your reason: then measure.

(Bisection of Angles.)

Draw an angle of any size on tracing paper, and fold it so as to bring one arm exactly over the other. Unfold your paper, and mark the crease. The crease **bisects** the angle, that is, divides it into two equal parts. Why?

How would you bisect an angle by means of your protractor?

Problem 3.

To bisect an angle AOB **with ruler and compasses.**

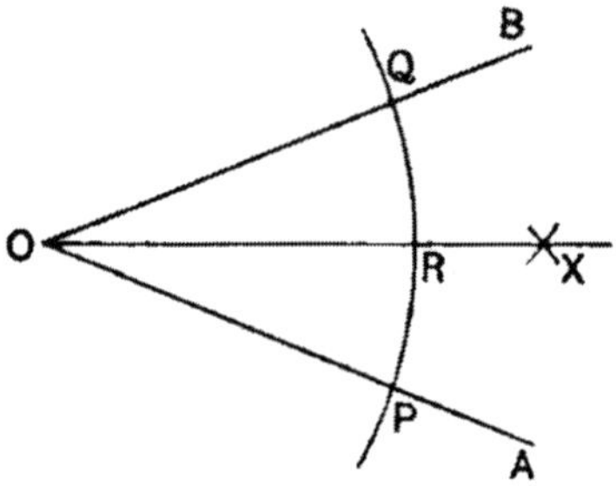

[The given angle AOB may be of any size: its arms may be conveniently taken about 9 cm. in length.]

Construction. With centre O, and any radius, draw an arc of a circle cutting OA at P and OB at Q.

Take in your compasses any length greater than half the distance from P to Q.

With centre P, and this length as radius, draw an arc. With centre Q, and *the same radius*, draw another arc, cutting the former at X.

Join OX.

Now, with your protractor, measure each of the angles AOX, BOX, and see if they are equal.

(Further Tests.)

(i) By means of tracing-paper, or by folding about OX, ascertain if the ∠AOX = the ∠BOX.

(ii) Put R where OX cuts the arc PQ. Compare with your dividers the distances (chords) RP, RQ. Are they equal? If so, how does this prove that the ∠s AOX, BOX are equal?

(iii) If the arc drawn from centre P had a greater radius than that drawn from centre Q, would OX still be the bisector of the ∠AOB? If not, towards which arm would OX lean?

You see then that OX is the bisector in our problem because we have worked from the arms OA and OB in exactly the same way.

(iv) In drawing the arcs from P and Q as centres, why did we take a radius *greater than half* PQ? What would have happened if the radius had been *less* than half PQ?

Ex. 9. With ruler and compasses only, draw an angle of 60°, and bisect it. Test your work with the protractor.

Ex. 10. Draw an angle of 150° with your protractor; then with ruler and compasses, divide the angle into *four* equal parts.

Ex. 11. With ruler and compasses construct an angle of 60°; then obtain from it an angle of 15°.

Ex. 12. Draw a straight line **AB**. In **AB** take a point **O**, and from it draw a line **OC** making any angle with **OA**. Bisect the angles **AOC**, **BOC** (*by construction*), and call the bisectors **OX** and **OY**. Measure the angle **XOY**. Can you account for the result?

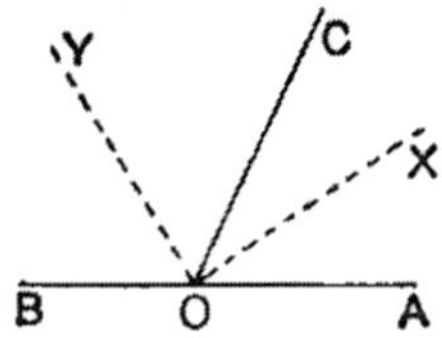

Ex. 13. Draw two straight lines **AB**, **CD** crossing one another at **O** at any angle. Bisect the angles **BOD**, **AOC** (*by construction*). Call the bisectors **OX** and **OY**. What do you notice as to the direction of these two bisectors?

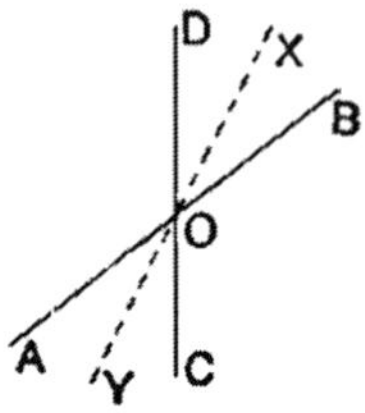

Ex. 14. Draw the patterns shewn below:

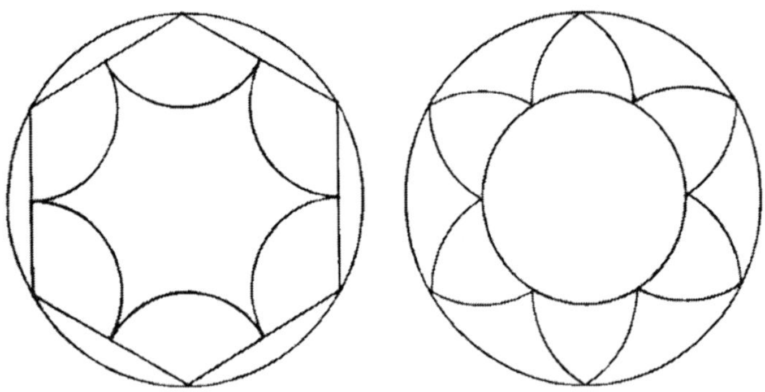

Fig. 1. Fig. 2.

(i) In Fig. 1. The circle is to be drawn first, radius 5 cm.

(ii) In Fig. 2. The *inner* circle is to be drawn first, radius 3 cm., the arcs of the star are to be of the same radius.

VII. DIRECTION. PARALLELS.

Suppose a man is walking along a straight path from A towards B; and suppose that, on reaching the point P, he alters his course, and proceeds along the path PC.

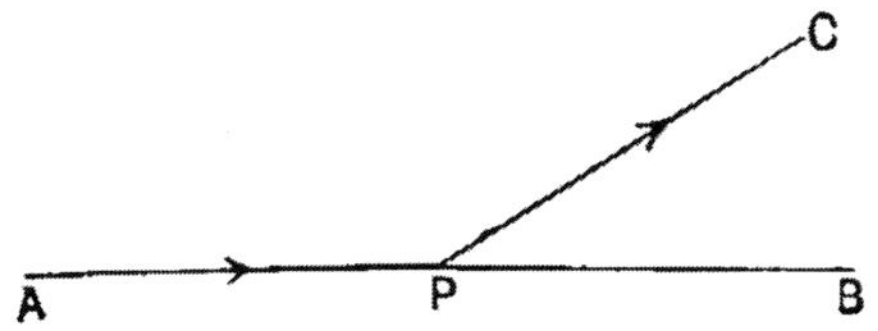

Then AB represents his first direction, PC his second direction; and his *change of direction* is given by the angle BPC, that is to say, the angle between his new course and *the line which he would have followed if he had gone straight on.*

Ex. 1. A ship sailing due East alters her course 25° towards North. Draw a diagram to represent this, marking the angle which shews her change of direction.

Ex. 2. A man walks due South, then turns 43° towards West. Shew by a Figure his first direction, his second direction, and his change of direction.

Two men are walking from A towards B. One on arriving at P changes his direction, say by 37°, to his left, following the line PC. The other goes straight on to Q, then also changes his direction by 37° to his left, following the line QD.

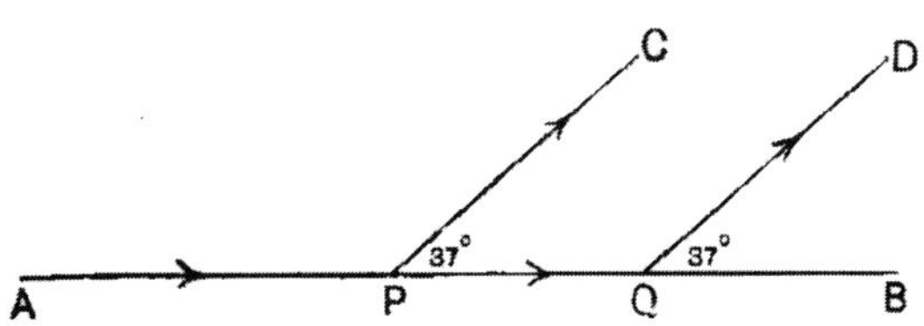

Do the two new paths meet? Does it seem to you that they would meet if they were prolonged ever so far forwards or backwards? Lines such as PC and QD, *which point in the same direction* never meet: they are said to be **parallel**; and the angles BPC, BQD, which fix the direction of these lines by comparison with AB, are called **corresponding angles.**

You have now to learn the use of the triangular rulers called **set squares**. Notice that one angle in each is a *right angle*: the remaining angles in one set square are both 45°; in the other they are 60° and 30°.

With a set square and a straight ruler we can draw parallel lines, as follows:

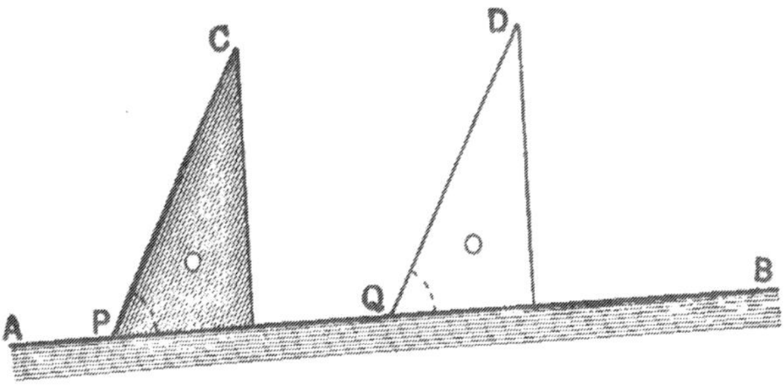

Place either set square in any position such as that shaded in the diagram, and against one of its sides lay a straight ruler (marked AB in the Figure). Holding the ruler firm, slide the set square along it, so that the side marked PC moves into the position QD. Then QD and PC are parallel. Why? Thus, if in any two positions of the set square we rule lines along the same edge, we get a pair of parallels.

Before going further practise yourself a little in this process, drawing pairs of parallel lines in various positions and directions. If the straight ruler has a bevelled edge, the set square is apt to slip up over it: in this case use the longest side of your other set square as a guide.

Ex. 3. Draw with your set squares two parallel lines AB, CD, about 10 cm. long and about 5 cm. apart. Draw *any* straight line EF across them, and number the angles so formed as in the diagram below.

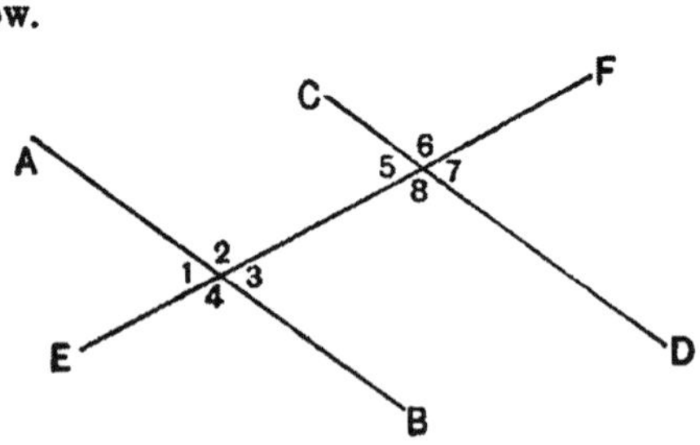

(i) Point out four pairs of **corresponding** angles.

Carefully measure each pair of corresponding angles with your protractor, and enter their values in your Figure. Having drawn EF at random your measurements shew that *corresponding angles* in each pair *are equal*. Note this.

(ii) The angles 3 and 5 are said to be **alternate**.

Point out another pair of alternate angles.

Looking back to your previous measurements, do you find alternate angles equal?

We may account for this by what has gone before, as follows:

The angle 5 = the angle 1. Why?

The angle 3 = the angle 1. Why?

Hence we see that the angle 5 must be equal to the angle 3.

(iii) The angles 3 and 8 are called **interior** angles.

Add together the angles 3 and 8.

Add together the angles 2 and 5.

Compare the results and try to account for them.

(iv) Put P and Q at the points where EF cuts the parallels AB, CD; then make a tracing of your figure. Move the tracing-paper so that the trace of EF slides along the original line EF, until the trace of P falls on Q. Where does the trace of AB fall? In this way verify the conclusions marked (i) and (iii).

Again slide your tracing-paper round until the trace of P falls on Q, and the trace of Q on P. Where do the traces of AB and CD fall? In this way verify the conclusion marked (ii).

Ex. 4. Draw two parallels AB, CD; and cut them by a line EF, making an angle of 57° with AB. Call this angle 1, and number the rest as before. Now write down (without measurement) the number of degrees in each of the angles 2, 3, 4, 5, 6, 7, 8.

Ex. 5. Repeat the last Exercise, drawing EF at an angle of 117° with AB. Write down (without measuring) the remaining angles.

Ex. 6. Draw a line AB about $3\frac{1}{2}''$ long. Take a point P about 2″ from AB. From A draw a line through P, and measure the angle PAB. Now, using your protractor, draw a line through P parallel to AB. Do this in two ways: (i) by making *corresponding* angles equal; (ii) by making *alternate* angles equal.

Problem 4.

Through a given point P *to draw* **with a set square** *a line parallel to a given straight line* AB.

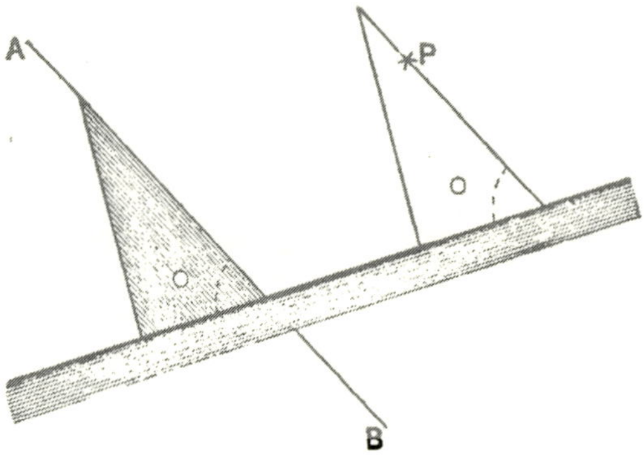

Place either set square so that one of its sides lies along AB in the position shaded in the diagram.

Against either of the other sides lay a straight-edge (either a straight ruler or the longest side of the second set square).

Then holding the straight-edge firmly, slide the set square along it until the side originally placed along AB passes through the point P.

A line ruled along this side is parallel to AB, for the *corresponding angles* marked in the diagram are necessarily equal.

Problem 5.

Through a given point P *to draw* **with ruler and compasses** *a line parallel to a given straight line* AB.

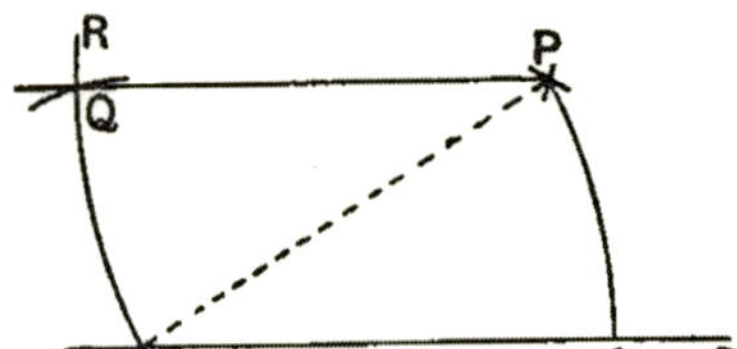

[A convenient figure is got by making AB about 8 cm. long, and placing P about 5 cm. from AB.]

Construction. With A (or any other point in AB) as centre, and the distance AP as radius, draw an arc cutting AB at C.

With P as centre, and the *same radius* PA, draw the arc AR.

Take the distance (or chord) PC in your compasses, and with centre A cut the arc AR at Q.

Join PQ.

Now test to see if PQ is parallel to AB.

(*Tests.*)

(i) Join AP, and ascertain by any practical means if the alternate angles CAP, QPA are equal. If so, AB and PQ are parallel.

(ii) Could you not conclude without measurement that the ∠CAP = the ∠QPA? Bear in mind that the arcs CP, QA have the same radius; that is, they are arcs of equal circles. Also remember that the arc AQ has been got by stepping off a chord equal to the chord PC. Now argue the rest out for yourself.

Ex. 7. Take two points A and B, 6 cm. apart. Through A draw any straight line; and through B draw a parallel line with your set squares.

Ex. 8. Draw a line AB of length 3″. With your protractor draw AC making an angle of 76° with AB. Now through B draw a line parallel to AC.

Do this Exercise twice, drawing the parallel (i) with set squares; (ii) with ruler and compasses.

Ex. 9. Repeat Ex. 8, making AB of length 9 cm., and the ∠BAC equal to 32°. Draw the parallel with your set squares; then test with ruler and compasses (by going through the construction of Problem 5).

Ex. 10. Draw a right angle AOB with your protractor, making each of the arms OA, OB 7·5 cm. in length. Through A draw a parallel to OB, and through B draw a parallel to OA. Do this with set squares.

What is the shape of the figure you have just drawn?

Ex. 11. Draw a straight line AB of length 7 cm. Find a point P that is 7 cm. from A and also 7 cm. from B. Through P draw a parallel to AB. (All this is to be done with ruler and compasses.)

Ex. 12. Draw a line AC, 2″ long, and bisect it by measurement at B. Through A, B, C draw parallels AX, BY, CZ in any direction (with set squares). Now draw any line across the parallels cutting them at P, Q, and R. Measure and compare PQ and QR.

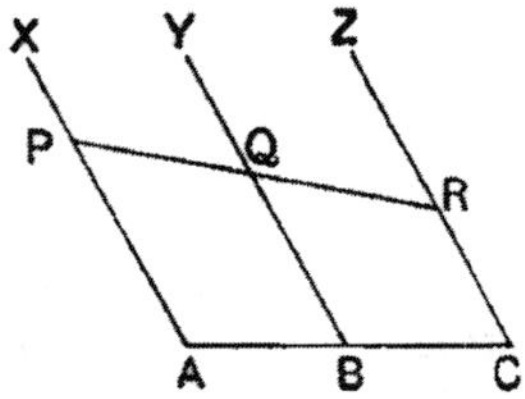

Draw any other line across the parallels cutting them at L, M, and N. Measure and compare LM and MN.

What conclusion do you arrive at from these experiments?

PROBLEM 6.

To divide a given straight line AB *into five equal parts (without measurement).*

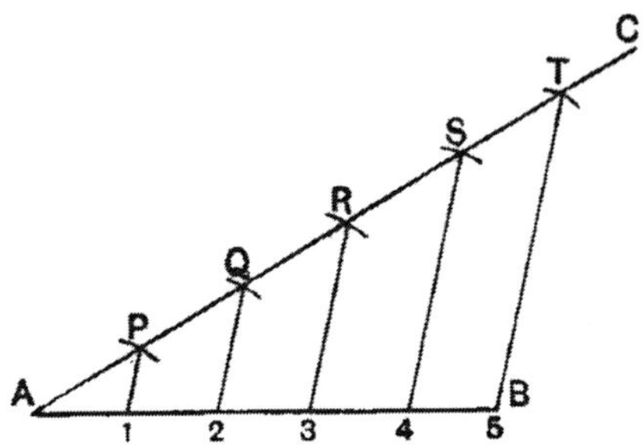

[The given line AB may be of any length; but do not measure it.]

Construction. From A draw AC, making any angle with AB.

Take any length in your compasses, and step it off *five* times along AC. Call the points of division P, Q, R, S, T.

Join TB.

Through P, Q, R, S draw parallels to TB (with set squares).

These parallels will divide AB into *five* equal parts. Test this with your dividers.

In the same way a straight line may be divided into *three*, *seven*, or any other number of equal parts.

The construction depends on the law which you will have found out from Ex. 12.

Ex. 13. Draw a line 2·7″ long, and divide it by the last construction into three equal parts. Test afterwards by measurement.

Ex. 14. Use the above method to *bisect* a line of 7·8 cm. Bisect the same line by Problem 1, p. 17, and see if your two results agree.

Ex. 15. Draw a line 3·2″ long, and divide it by the above construction into *four* equal parts.

How else could you divide a line into four equal parts, using ruler and compasses only?

Ex. 16. Draw a line 9 cm. long, and divide it into *seven* equal parts. Test with your dividers.

Ex. 17. From a line 3·5″ long, cut off *one-fifth* part by construction.

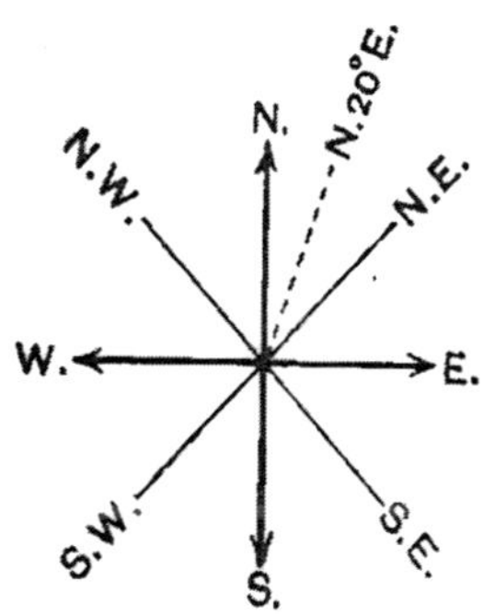

The line of direction which bisects the angle between North and East, is called *North-East*; and the terms North-West, South-East, South-West, have corresponding meanings.

If, looking from a light-house, a ship is seen in the direction North-West, we say that it *bears* N.W. from the light-house, or that its *bearing* is N.W. If the direction of the ship, as seen from the light-house, makes with the line pointing North an angle of 20° on the East side of that line, we say that the ship bears 20° East of North, or N. 20° E.

Ex. 18. A man walks 6 kilometres due East, then 5 kilometres due North. Draw a plan (scale 1 km. to 1 cm.), and find by measurement how far he is from his starting-point.

Ex. 19. North West from my garden gate is a cottage, 300 yards distant: North East of the cottage and 250 yards from it is a well. Draw a plan (scale 100 yards to 1 inch), and find as nearly as you can how far the well is from the garden gate.

Ex. 20. Two cyclists, each riding 14 km. an hour, leave a house at the same time. One goes by a straight road leading S.E.; the other by a road leading S.W. How far apart will they be in half an hour? (Scale 1 km. to 1 cm.)

Ex. 21. A man goes South 4 miles, then West 6 miles, then South again 4 miles. How far is he now from his starting-point? (Scale 2 miles to 1 inch.)

Ex. 22. A ship on leaving port sails N.W. for 18 miles, then North for 15 miles. Shew her course on the scale of 10 miles to 1 inch. Find her approximate distance, and her bearing from the port, that is, how many degrees West of North.

Ex. 23. A boy walks 200 yards in a certain direction; then, turning 68° to his left, he walks 300 yards; finally he turns 68° to his right, and walks 250 yards. Shew his track on a plan (100 yards to 1 inch); and explain why his third direction is parallel to his first. How far is he at last from his starting-point?

Ex. 24. A traveller wishes to go due North, but finds his way barred by a swamp. He therefore walks 5 kilometres N.E., then 5 kilometres North, then 5 kilometres N.W.; and now he finds himself due North of his starting-point. How many kilometres has he lost by having gone out of his way? (Scale 1 km. to 1 cm.)

Ex. 25. Draw the patterns given below: the dimensions should be twice those of the copies.

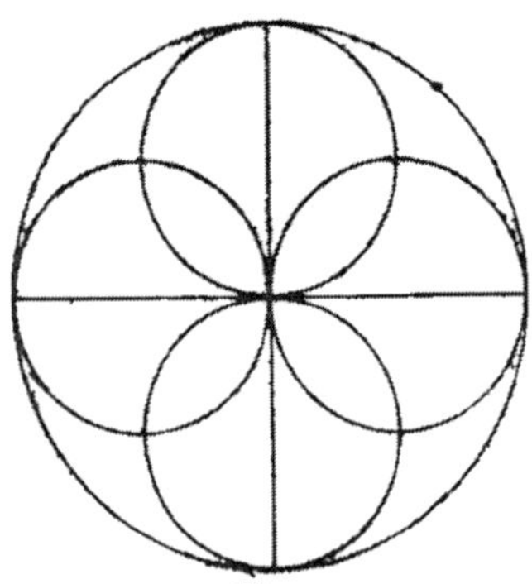

Fig. 1.

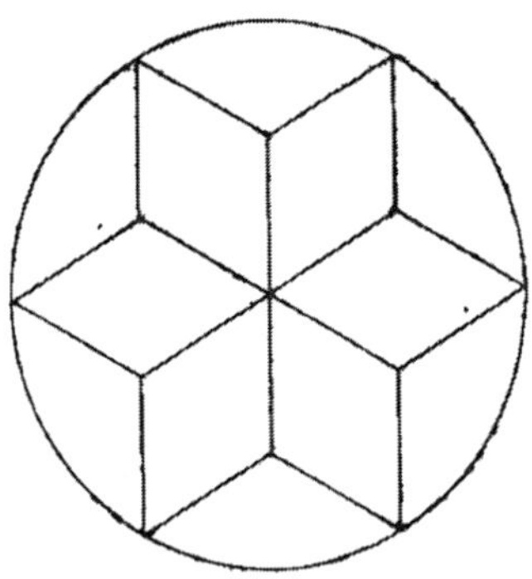

Fig. 2.

VIII. PERPENDICULARS.

PROBLEM 7.

To draw **with ruler and compasses** *a straight line bisecting a given straight line* AB **at right angles.**

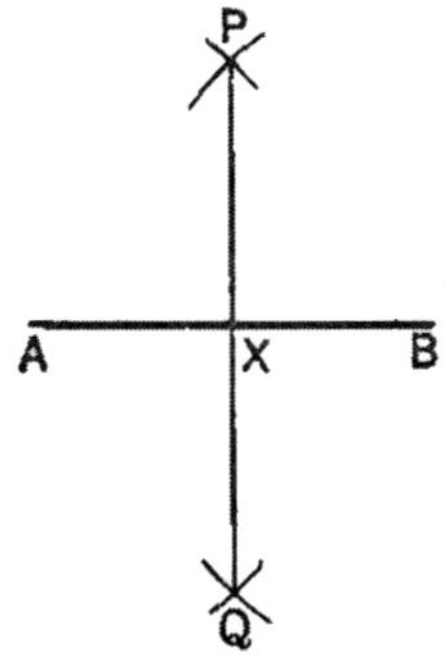

Construction. Follow the method of Problem 1, p. 17.

Your experiments have already shewn that PQ cuts AB at its middle point. We have now to satisfy ourselves that PQ is at right angles, or perpendicular to AB. Test this first with your protractor.

(*Further Tests.*)

(i) Make a tracing of your figure, and fold it so as to bring A over B. Note the position of the crease, and explain the result.

(ii) Make a tracing as before, and turn it about the point X until the trace of XA lies along XP. Where does the trace of XP fall? Shew from this that PQ is at right angles to AB.

Ex. 1. Draw a straight line AB, 8 cm. long, and bisect it at right angles by a line PQ. Use radii of length 5 cm.; and measure PQ.

Ex. 2. Draw a line of any length, say 2·4″, on tracing paper; and bisect it at right angles by PQ, choosing your own radii for the arcs.

Shew by measurement that PQ bisects AB, *and also that* AB *bisects* PQ.

If you fold the figure about PQ where will the point A fall?

If you fold the figure about AB where will the point P fall?

The figure is *symmetrical* about PQ; and also about AB.

Ex. 3. Take a line AB of length 7 cm. With centre A and radius 5 cm. draw a circle. With centre B and radius 4 cm. draw a circle cutting the first at P and Q. Join PQ. "*Then* PQ *bisects* AB *at right angles.*" Which part of this statement is true? Which part is false?

Ex. 4. Draw a line AB of any length you like, and bisect it at right angles by PQ, choosing the radii yourself: note the length of the radii you use.

How far is P from A and B? How far is Q from A and B?

Take *any* point in PQ, and call it R. Measure RA and RB, and compare their lengths. You will find that R is *equidistant* from A and B.

Ex. 5. Draw a line AB, 8 cm. long. With your compasses find two points P and Q distant 7 cm. from both A and B; also two points R and S distant 6 cm. from both A and B; also two points X and Y distant 5 cm. from both A and B. On what line do all these points lie? How many points are there which are 4 cm. from both A and B?

PROBLEM 8.

Through a given point P *to draw* **with set squares** *a line* **perpendicular** *to a given straight line* AB.

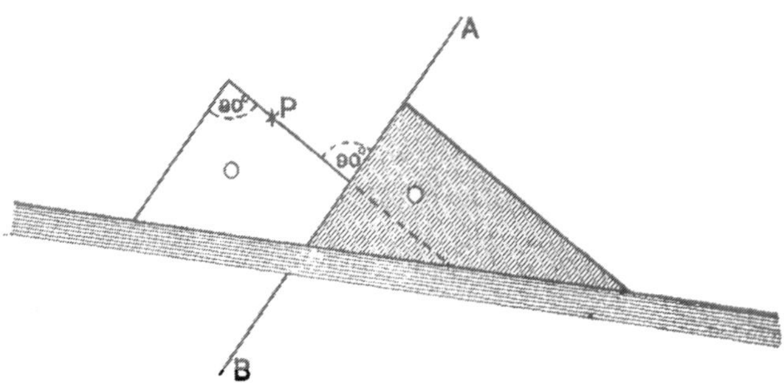

Take either set square and place one of the sides containing the right angle along AB.

Apply the straight-edge to the longest side (*i.e.* the side opposite the right angle) of the set square; and slide the latter until the side originally perpendicular to AB passes through P.

A line ruled along this side will be perpendicular to AB, for the alternate angles marked in the diagram are equal.

NOTE. Following the principle of this method, you should devise for yourself arrangements of a set square and straight edge by which a line may be drawn through a given point P making with a given line AB an angle (i) of 45°, (ii) of 60°, (iii) of 30°.

(Exercises to be done with Set Squares.)

Ex. 6. Draw a straight line AX, and mark off along it AB, BC, CD, each 1″ in length. Through A, B, C, and D draw lines perpendicular to AX. Why are these lines parallel?

Ex. 7. Draw a line AB of length 7 cm. Through A draw a perpendicular to AB, and along it measure AC 7 cm. long. Through B draw a parallel to AC; and through C draw a parallel to AB.

What is the shape of the figure you have thus drawn?

Ex. 8. Draw a line AB, 8 cm. long. Draw AC perpendicular to AB, and make AC=6 cm. Join BC. From A draw AD perpendicular to AC. Measure AD.

Ex. 9. Draw a line AB. At A make (with your set squares) (i) a right angle, (ii) an angle of 60°, (iii) an angle of 30°, (iv) an angle of 45°.

Ex. 10. (*a*) Draw a line AB of length 10·5 cm. From A (with your protractor) draw AC and AD making angles of 45° with AB, one on each side. Through B draw (with set squares) parallels to AC and AD.

What is the shape of the figure you have thus drawn?

(*b*) Take a line AB, 6 cm. long. Using a radius also of 6 cm., bisect AB at right angles by PQ. From centre X (the point of bisection) with radius 3 cm. mark off points C and D in PQ. Join CA, CB, DA, DB.

What is the shape of the figure ACBD? Measure the angles DAC, ACB.

(*c*) Make a straight line BC of length 7·6 cm.; and through B and C draw two parallels BP, CQ, (with set squares) in any direction. Bisect the angles PBC, QCB by lines meeting at A, and measure the angle BAC.

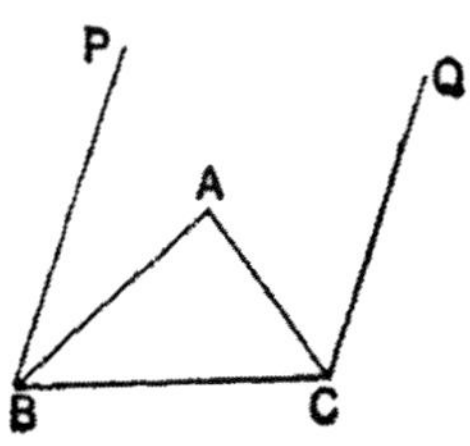

Problem 9. First Method.

To draw, **with ruler and compasses,** *a straight line* **perpendicular** *to a given straight line* AB *at a given point* X *in it.*

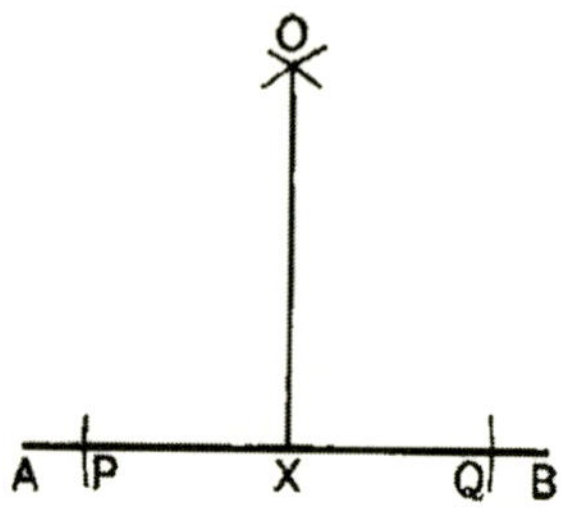

[The given straight line AB may be of any length, for convenience say about 4″. The given point X in this construction should not be taken near an end of AB: take X about 1·5″ from A.]

Construction. Take in your compasses any length less than XA (say a little over 1″), and with X as centre mark off two points P and Q in AB.

Now take in your compasses any length greater than PX (say about 2″), and first with P as centre, then with Q as centre, draw two arcs cutting at O.

Join OX.

Now test with your protractor to see if OX is perpendicular to AB.

(*Further Tests.*)

(i) Test with a set square and straight-edge as explained in Problem 8, p. 42.

(ii) Use the test marked (ii) in Problem 7, p. 41.

(iii) Invent a test with ruler and compasses to find if the angles OXP, OXQ are equal. (See Problem 2, p. 28.) If they are, how does this shew that OX is perpendicular to AB?

PROBLEM 9. SECOND METHOD.

[When the given point X is at or near one end of AB.]

Construction. With X as centre, and any length as radius, draw the arc CDE, cutting AB at C.

With the *same radius* step off from C the points D and E round the arc.

With D and E as centres, and the same radius as before, draw arcs cutting at O.

Join XO.

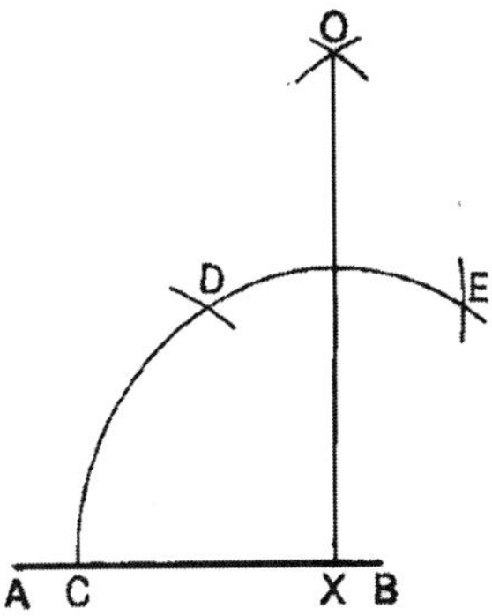

(*Verification.*)

Join DX and EX.

How many degrees are there in the ∠s CXD, DXE? Why? [p. 18.]

How many degrees are there in the ∠s DXO, EXO? Why? [p. 30.]

How many degrees are there in the ∠AXO?

(*Perpendiculars by Construction.*)

Ex. 11. Draw a straight line 4″ long. At points $1\frac{1}{2}$″ from each end erect perpendiculars (First Method). Why are these parallel?

Ex. 12. Draw a line AB, 6 cm. long. At each end erect perpendiculars AC, BD (Second Method), each 6 cm. long. Join CD.

Name any test by which you can find if CD is parallel to AB.

Ex. 13. Employ Problem 9 (Second Method) and Problem 3 (p. 30) to draw lines making with a given line AB angles of 90°, 45°, $22\frac{1}{2}$°.

Ex. 14. By constructions with ruler and compasses draw lines making angles of 60°, 30°, 15° with a given line AB.

How would you draw a *reflex* angle of (i) 270°, (ii) 300°?

Ex. 15. Construct a perpendicular at the end A of a given line AB. Then, with ruler and compasses, draw AC making an angle of 135° with AB.

PROBLEM 10.

To draw **with ruler and compasses** *a straight line* **perpendicular** *to a given line* AB *from a given point* X *outside it.*

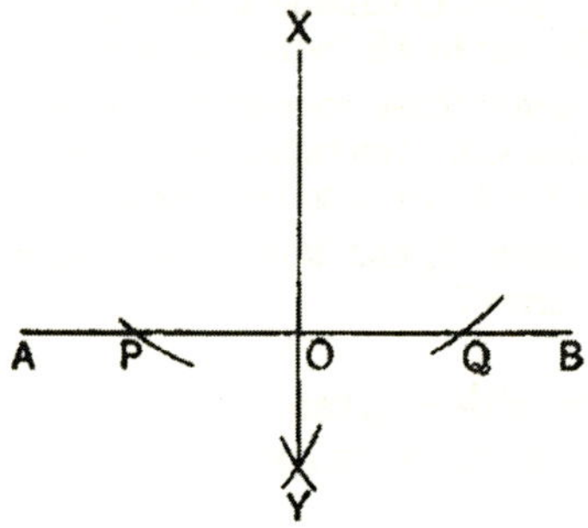

[The given line AB may be taken about 4″ long.]

Construction. With centre X, and any radius of sufficient length, cut AB at P and Q.

Take in your compasses any length greater than half PQ.

With centre P, and this length as radius, draw an arc on the side of AB opposite X;

With centre Q, and the *same radius*, draw an arc cutting the last arc at Y.

Join XY, cutting AB at O.

Now apply any of the tests previously explained to ascertain if XO is perpendicular to AB.

Ex. 16. With your set square or protractor draw a right angle AOB; and make OA=7·5 cm., and OB=5·5 cm. Join AB, and by construction drop a perpendicular on AB from O.

Ex. 17. Draw a line AB of length 1·6″. Find with your compasses a point P distant 1·7″ from both A and B. From P drop a perpendicular PM on AB (by construction). Measure PM.

Ex. 18. Draw a straight line AB, and take any point P outside it. Draw PX perpendicular to AB (with set squares). Measure PX.

Now take any two points Y, Z in AB on the same side of X. Join and measure PY, PZ.

Of the lines PX, PY, PZ, which is least? Which is greatest? Can you draw from P to AB a shorter line than the perpendicular PX?

The distance of a point P from a straight line AB is understood to be the *length of the perpendicular* PX, this being the shortest line that can be drawn from P to AB.

Ex. 19. Take a point O outside a straight line AB, and from O draw OX perpendicular to AB (with set squares).

With centre O draw three concentric circles: the radius of the first is to be *less than* OX; the radius of the second is to be *equal to* OX; the radius of the third is to be greater than OX.

Now carefully notice if, and how, these circles meet AB. What conclusion do you draw?

A circle drawn with a given point O as centre will *touch* a given line AB if its radius is equal to the perpendicular from O to AB.

If the radius is greater than this perpendicular, the circle will cut AB in two points; if less, the circle will not meet AB at all.

Ex. 20. Draw the patterns shewn below. Your drawings should be twice the size of the copies.

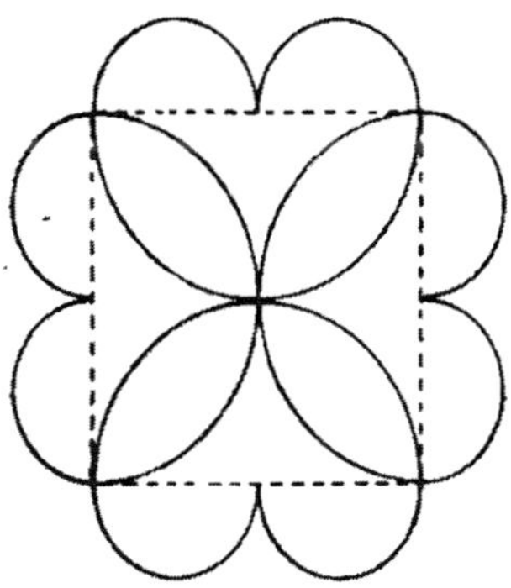

Fig. 1.

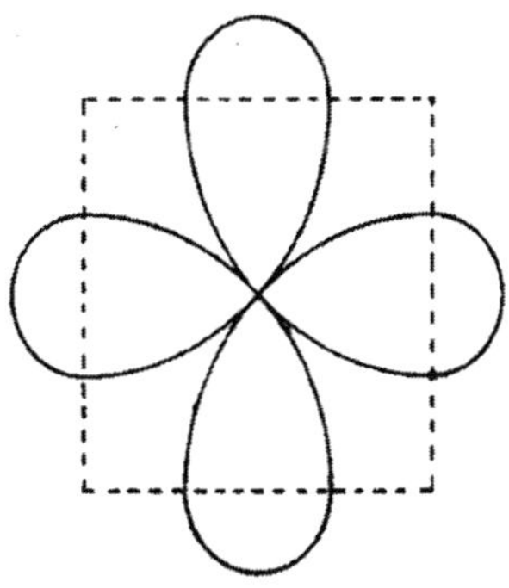

Fig. 2.

In Fig. 1. the square is drawn first, and the curves are all semi-circles.

In Fig. 2. the square is drawn first. The inner arcs are drawn from the vertices of the square as centres, and half the diagonal as radius: the other curves are semi-circles.

IX. TRIANGLES.

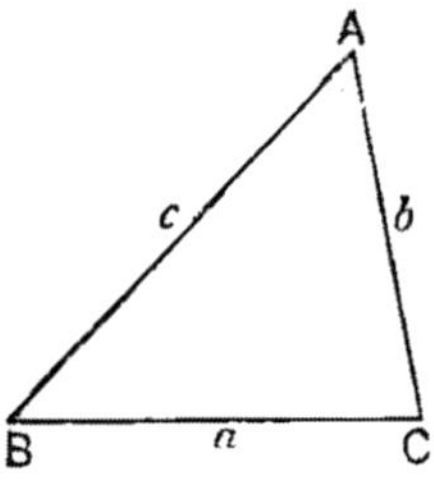

Take any three points A, B, and C not all in a straight line, and join AB, BC, CA. The figure thus formed is called a **triangle**: it has three vertices, three sides, and three angles.

The letters A, B, C are used not only to name the vertices, but to represent the size of the corresponding angles as measured in degrees; while a, b, c are taken to represent the lengths of the opposite sides.

Thus in the Figure $\begin{cases} A = 58°, & B = 44°, & C = 78°; \\ a = 2{\cdot}6 \text{ cm.}, & b = 2{\cdot}1 \text{ cm.}, & c = 3{\cdot}0 \text{ cm.} \end{cases}$

The symbol △ is used as an abbreviation for the word *triangle.*

Suppose the Figure represents a triangular field, and you wish to walk from the corner B to the corner C. Which would be the longer way, to go from B to A and then from A to C, or to go straight from B to C along the side BC?

Which is the greater, AB + BC, or AC?

Which is the greater, BC + CA, or BA?

You see at once that *any two sides of a triangle must be together greater than the third side.* Indeed we have already seen the truth of this; for in the first chapter we observed that the *straight line joining two points is the shortest distance between them.*

Ex. 1. To illustrate this further, draw any triangle **ABC**; measure its sides, and fill up the following form:

$a+b=$	cm.	$b+c=$	cm.	$c+a=$	cm.
$c=$	cm.	$a=$	cm.	$b=$	cm.
Excess =	cm.	Excess =	cm.	Excess =	cm.

PROBLEM 11.

To draw a triangle, having given the **three sides.**
(For instance: $a=8$ cm., $b=7$ cm., $c=6$ cm.)

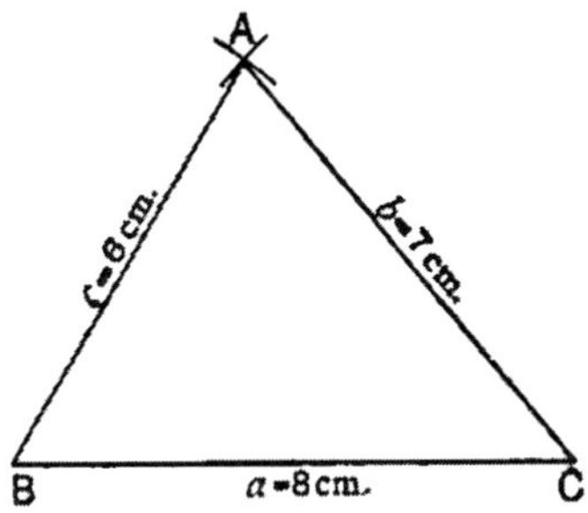

Construction. Draw a straight line BC of length 8 cm.

With centre B, and a radius of 6 cm. (the length of c), draw a circle.

With centre C, and a radius of 7 cm. (the length of b), draw a second circle cutting the first at A.

(*Arcs* of these circles, shewing the cutting point, are enough in practice.)

Join AB and AC.

Then ABC is the triangle required.

(Remarks.)

(i) Notice that the problem is the same as that of finding a point A distant 6 cm. from B, and 7 cm. from C. Can more than one such point be found?

(ii) Draw *two* triangles, one on each side of BC, having the dimensions given above.

Cut out the double figure so formed, and fold it about BC. What do you find? Are the two triangles of the same size and shape?

(iii) Go through the construction of Problem 11 with the following dimensions: $a=8$ cm., $b=4$ cm., $c=3$ cm.

What difficulty arises? Why is the construction impossible?

(iv) Go through the construction with these dimensions: $a=8$ cm., $b=5$ cm., $c=3$ cm.

Observe carefully what happens, and give a reason for it. Can you draw a triangle whose sides have these lengths?

Ex. 2. Construct (or try to construct) triangles whose sides have the following lengths.

If any set of lengths seems to you impossible, carry out the construction as far as it can go, and then say how and why it fails.

(i)	$a=3{\cdot}0''$,	$b=3{\cdot}0''$,	$c=3{\cdot}0''$.
(ii)	$a=3{\cdot}0''$,	$b=2{\cdot}5''$,	$c=2{\cdot}5''$.
(iii)	$a=3{\cdot}0''$,	$b=2{\cdot}5''$,	$c=2{\cdot}0''$.
(iv)	$a=3{\cdot}0''$,	$b=1{\cdot}5''$,	$c=1{\cdot}0''$.
(v)	$a=3{\cdot}0''$,	$b=2{\cdot}0''$,	$c=1{\cdot}0''$.
(vi)	$a=5{\cdot}4$ cm.,	$b=7{\cdot}6$ cm.,	$c=5{\cdot}4$ cm.
(vii)	$a=4{\cdot}5$ cm.,	$b=7{\cdot}0$ cm.,	$c=3{\cdot}5$ cm.
(viii)	$a=4{\cdot}5$ cm.,	$b=7{\cdot}0$ cm.,	$c=2{\cdot}0$ cm.
(ix)	$a=4{\cdot}5$ cm.,	$b=7{\cdot}0$ cm.,	$c=2{\cdot}5$ cm.
(x)	$a=5{\cdot}4$ cm.,	$b=8{\cdot}2$ cm.,	$c=4{\cdot}3$ cm.

In each of the above triangles, when possible, measure all the angles very carefully, and enter the measurements on your drawings: these measurements will be wanted later on.

If a triangle has *all* its sides equal, it is said to be **equilateral**;
if it has *two* sides equal, it is called **isosceles**;
if none of the sides are equal, it is called **scalene**.

Ex. 3. Point out examples from the triangles you have just drawn of equilateral, isosceles, and scalene triangles. Notice their shapes carefully.

In an isosceles triangle *the* vertex is usually understood to be the point at which the *equal* sides meet; then the opposite side is called the **base**.

(Comparison of Sides and Angles.)

Ex. 4. Measure the angles of the equilateral triangle you have drawn in Ex. 2 (i). Are the angles equal? How many degrees are there in each?

Draw any larger or smaller equilateral triangle, and measure its angles. Do you find that equilateral triangles of different sizes have the same shape?

Ex. 5. Take the isosceles triangle you have drawn in Ex. 2 (ii). Measure and compare the *base angles*, namely those at B and C, which are opposite to the equal sides.

Draw *any* isosceles triangle ABC (A at the vertex) without measuring the sides: measure and compare the angles at the base.

Make a tracing of your triangle; turn the tracing over, and see if it can thus be fitted over the original triangle ABC. If so, where does the trace of the ∠B fall? And where does the trace of the ∠C fall?

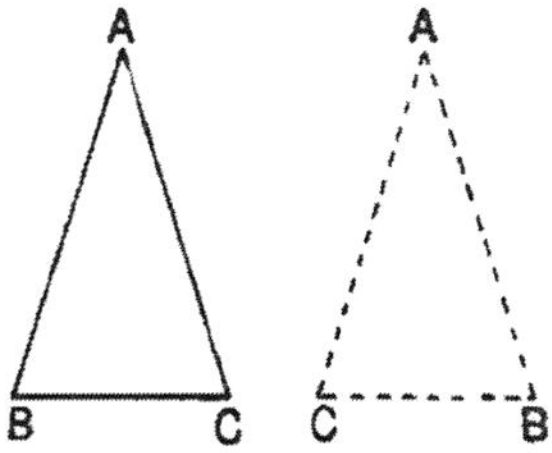

Now state the conclusion you draw from these experiments.

Lastly bisect the angle BAC, in your tracing, and fold the figure about the bisector. How does this experiment support your conclusion?

Ex. 6. Take the scalene triangle you have drawn in Ex. 2 (iii). Measure the angles. Which is the greatest side, and which is the greatest angle? Which is the smallest side, and which is the smallest angle?

Draw a triangle of any size and shape you like (not from measurements). Now measure the sides and angles. Write down the sides a, b, c in order of their lengths, beginning with the longest. Write down the angles A, B, C in order of their size, beginning with the largest.

State in your own words the conclusion you draw.

(i) In each of the six possible triangles given in Ex. 2 you have measured the angles; let us in each case add the three angles together, and write down the result thus:

$$A + B + C = \quad \text{degrees.}$$

Range the totals in a column; compare them carefully, always bearing in mind that there may be small errors in your measurements. Take the average.

(ii) Now draw any three triangles varying in size and shape (not from given measurements). Measure the angles in each case, and add them together. Compare the *sums.*

(iii) Draw a good sized triangle of any shape you like. Cut it out and tear off the corners. Fit these together at a point O; and observe the two outer straight edges. Do these fall in a straight line? If so, what do you learn from this experiment?

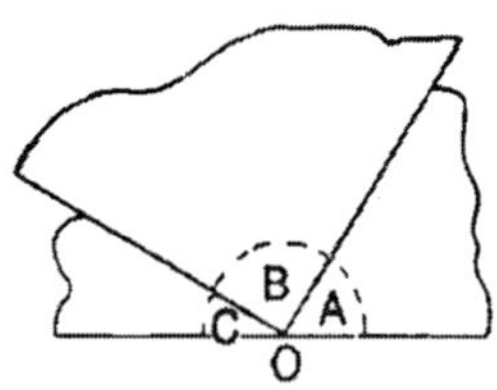

You have now reason for believing that in *any* triangle ABC

$$A + B + C = 180°;$$

or, in words, *the sum of the three angles is equal to* **two right angles**.

A triangle is said to be **right-angled** when *one* of its angles is a right angle.

Ex. 7. Draw a right-angled triangle. Can a triangle have more than one right angle? Can a right-angled triangle also have an obtuse angle? How many acute angles has a right-angled triangle?

A triangle is said to be **obtuse-angled** when *one* of its angles is obtuse.

Ex. 8. Draw an obtuse-angled triangle. How many acute angles must every obtuse-angled triangle have?

A triangle is **acute-angled** when *all three* of its angles are acute.

Ex. 9. Draw an acute-angled triangle. Why would it not be enough to say "A triangle is acute-angled when *one* of its angles is acute?"

Before constructing a triangle or other figure having given sides and angles, it is very useful to draw a rough free-hand sketch, in order to make sure that the question is understood, and to shew what is given and what is required.

PROBLEM 12.

To draw a triangle having given **two sides** *and the* **included angle.** (For instance: $b = 1{\cdot}8''$, $c = 2{\cdot}7''$, $A = 65°$.)

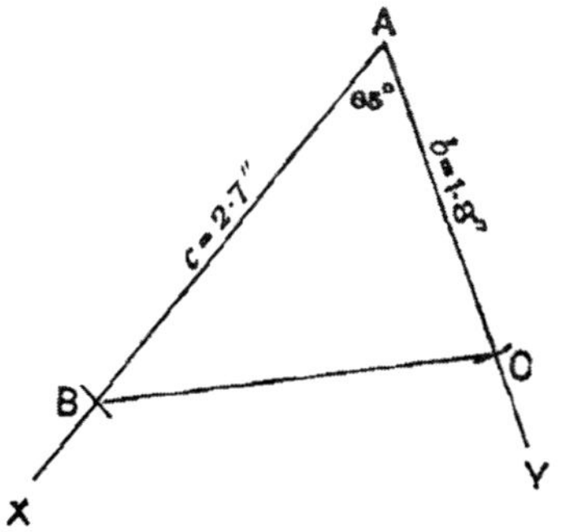

Construction. Draw a line AX; and from A draw AY making an angle of 65° with AX (using protractor).

From AX cut off AB equal to $2{\cdot}7''$ (the length of c).

From AY cut off AC equal to $1{\cdot}8''$ (the length of b).

Join BC.

Then ABC is evidently the required triangle.

[Measure the $\angle^s$ at B and C, and verify $A + B + C = 180°$.]

Ex. 10. Draw a right angle BAC (with protractor or set square), making AB and AC each $2{\cdot}5''$. Join BC.

Why are the angles at B and C equal? How many degrees are there in each?

Ex. 11. Draw a triangle in which $b = 7{\cdot}8$ cm., $c = 6{\cdot}2$ cm., and $A = 118°$. Measure a, B, and C; and verify $A + B + C = 180°$.

Ex. 12. Draw an isosceles triangle ABC, in which

$$AB = AC = 7{\cdot}0 \text{ cm., and } A = 84°.$$

Can you tell without measurement how many degrees there must be in each of the angles at B and C?

PROBLEM 13.

To draw a triangle having given **one side** *and the* **two angles** *at its ends.* (For instance: $a = 3{\cdot}2''$, $B = 72°$, $C = 37°$.)

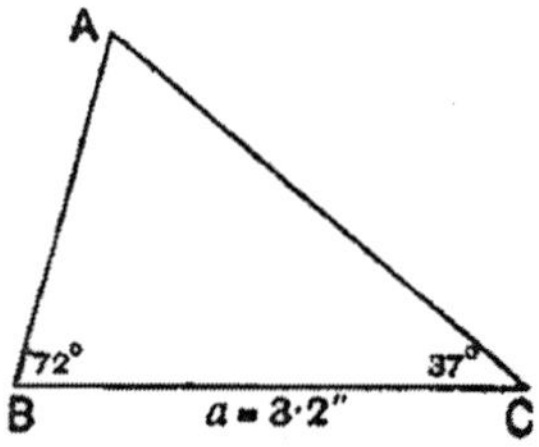

Construction. Draw BC equal to 3·2″.

At B make an angle of 72° with BC (using protractor).

At C make an angle of 37° with CB, on the same side as before.

Produce (that is to say, *prolong*) the lines to meet at A.

Then ABC is the required triangle.

[State, without measuring, the size of the angle A: then test your answer with the protractor.]

(Comparison of Angles and Sides.)

Ex. 13. Draw a triangle ABC in which $a = 6{\cdot}8$ cm., $B = 101°$, and $C = 44°$. Say, before drawing, what must be the size of the angle A. Verify afterwards by measurement.

Ex. 14. Each of the angles at the base of a triangle is 65°; what is the vertical angle?

Draw a triangle ABC in which $a = 2{\cdot}4''$, $B = C = 65°$. Measure b and c, and say what kind of triangle it is (i) in respect of its sides (ii) in respect of its angles.

Ex. 15. Draw a triangle ABC in which $b = 6{\cdot}2$ cm., $A = 61°$, and $C = 35°$. What is the angle B? Measure a and c.

Write down (i) the sides, (ii) the angles in order of their size, and compare the two results.

Ex. 16. Try to draw triangles in which

(i) $a=5{\cdot}8$ cm., $B=110°$, $C=70°$;

(ii) $a=5{\cdot}8$ cm., $B=45°$, $C=135°$.

What difficulty arises? Perhaps you find that the other sides would not meet on your paper: would they *ever* meet? Give a reason for your answer.

Ex. 17. In a right-angled triangle, if one acute angle is 60°, what is the other?

Draw a triangle in which $a=3{\cdot}0''$, $A=90°$, $B=60°$.

X. TRIANGLES CONTINUED. CONGRUENCE. PRACTICAL APPLICATIONS.

If you look back at Problems 11, 12, and 13, on the construction of triangles, you will notice that in each case *three* things were given: namely

(i) Three sides. (Problem 11);

(ii) Two sides and the included angle. (Problem 12);

(iii) One side and two angles. (Problem 13);

And these *data* (or things given) were enough to fix the size and shape of the triangle.

Ex. 1. Draw a good-sized triangle **ABC** of any shape; then state three different methods (corresponding to Problems 11, 12, and 13) by which an exact copy of it may be made.

Make a copy of the given triangle **ABC** in each of these ways; and test by seeing if a tracing of the triangle **ABC** can be exactly fitted over each copy.

Would the size and shape of a triangle be fixed if we were given the *three angles*? First of all, the sum of the three angles must be 180°, otherwise no triangle could be drawn from them.

Let us take A = 55°, B = 80°, C = 45°. Draw a line BC of *any* length as base. Make the angle B equal to 80°, and the angle C equal to 45°; then *whatever length we take for the base* BC, the third angle A must be 55°.

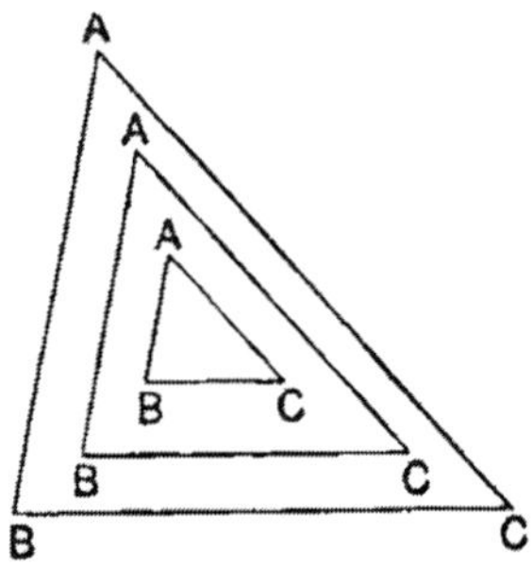

Thus any number of triangles of *different sizes* can be drawn having the given angles 55°, 80°, 45°. You will easily see that all these triangles have the *same shape*: in fact the three angles fix the *shape* but not the size of a triangle.

If a tracing of one triangle can be made to fit exactly over another, it is clear that the two triangles have the same size and shape, and are equal in all respects. The fitting of one figure over another for the purpose of comparison is called *superposition*; and if one figure *exactly* fits over the other, it is said to *coincide* with it. Figures which can be made to coincide with one another, thus shewing that they have the same size and shape, are said to be *congruent*.

(*Questions to be answered orally.*)

Ex. 2. In a △ABC, A = 70°, C = 50°; what is B?

Ex. 3. In a △ABC, B = 28°, C = 112°; what is A?

Ex. 4. How many triangles can there be in which A = 91°, B = 35°, C = 54°?

Ex. 5. How many triangles can there be in which A = 115° B = 50°, C = 25°?

Ex. 6. A △ ABC is right-angled at A; if B = 55°, what is C?

Ex. 7. In a △ ABC, B = 65°, and C = 25°. What sort of triangle is it (i) in respect of its angles, (ii) in respect of its sides?

Ex. 8. The △ ABC is isosceles, A being the vertex. If B = 41°, what are the other angles?

Ex. 9. The △ ABC is isosceles, and the vertical angle A is 50°. What are the angles at B and C?

Ex. 10. The isosceles △ ABC is right-angled at the vertex A. What are the angles B and C?

Ex. 11. In a △ ABC, if A + B = C, what is the angle C?

(*Exercises in Geometrical Drawing. The constructions to be done with ruler and compasses only unless otherwise stated.*)

Ex. 12. Draw a line AB of length 6 cm. Construct two equilateral triangles APB, AQB on opposite sides of AB as base.

Compare your construction with that of Problem 7 (p. 41), and explain why PQ bisects AB at right angles.

Ex. 13. On a base of 2·0″ draw an isosceles triangle, each of the equal sides being 2·5″.

From the vertex draw a perpendicular to the base; and shew by measurement that this perpendicular bisects the vertical angle. Account for this by comparing the constructions of Problem 10 (p. 46) and Problem 3 (p. 30).

Ex. 14. Draw a triangle ABC in which a = 7·6 cm., B = 80°, C = 46°. (With protractor.)

Bisect the ∠ BAC (construction) by a line which meets the base at X. Calculate the ∠ˢ AXB, AXC; and verify by measurement.

Ex. 15. On a base BC of 8 cm. construct an isosceles triangle ABC, having the angle at each end of the base half a right angle. (With protractor.)

Bisect BC at right angles by a line PQ. Why does PQ pass through A?

Ex. 16. Construct a triangle, having given: a = 5 cm., B = 60°, C = 90°. (Without protractor.) What is the ∠ A?

Ex. 17. Construct an angle **BAC** of 120°. Make **AB** = **AC** = 7·2 cm. Join **BC**.

What are the angles at **B** and **C**? Measure **BC** to the nearest millimetre.

Ex. 18. Construct (with ruler and compasses only) a triangle **ABC**, having the two sides **AB**, **AC** equal to two given lines c and b, and the included angle **A** equal to a given angle **P**.

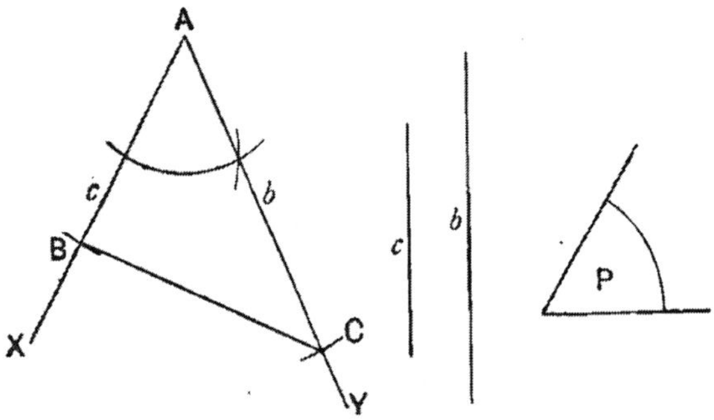

[This is Problem 12 set in a new form. We give the complete figure, and leave the details of construction to the pupil.]

Ex. 19. Construct (without protractor) a triangle **ABC**, having the side **BC** equal to the line a, and the angles **B** and **C** equal to the given angles **P** and **Q**.

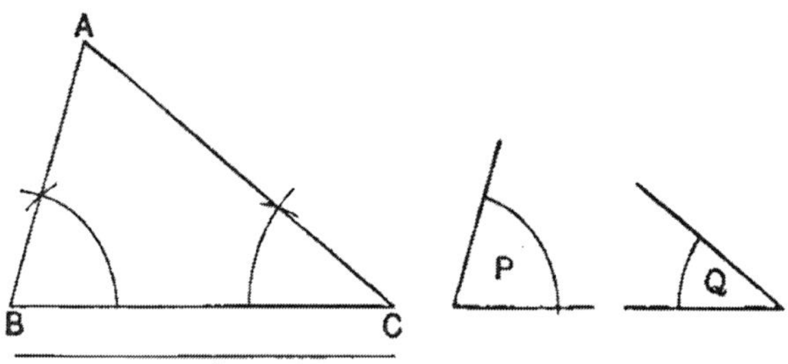

[This is Problem 13: as before, we leave the construction to the pupil. The teacher should furnish data for practice in the constructions of Exercises 18 and 19.]

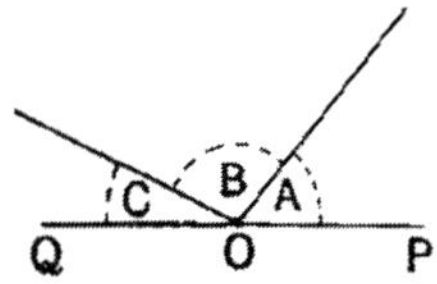

Ex. 20. Draw a straight line PQ of any length, and take a point O in it. From O draw two lines on the same side of PQ, and call the angles so formed A, B, and C.

On a base of 3·4″ construct a triangle having the angles at each end of the base equal to B and C. How do you know that the third angle of this triangle must be equal to A?

Ex. 21. Draw a triangle ABC having sides 10 cm., 9 cm., 8 cm. in length.

Bisect each side at right angles (Problem 7). If your drawing is correct, the bisectors meet at a point. Call the meeting point O.

Measure the distances of O from A, B, and C. Can you account for these distances being the same?

From centre O, with radius OA, draw a circle: this should pass through B and C.

A circle which passes through all the vertices of a figure is said to be **circumscribed** about it.

Ex. 22. Construct a triangle in which $a=2{\cdot}7''$, $b=3{\cdot}0''$, and $c=2{\cdot}3''$; and draw a circle to pass through its vertices by the method of Ex. 21.

Ex. 23. Construct an equilateral triangle on a base of 7 cm., and circumscribe a circle about it.

Ex. 24. Draw a good sized triangle of any shape. Through each vertex draw a line perpendicular to the opposite side (with set squares). What do you notice with regard to the meeting of the three perpendiculars?

Ex. 25. Draw a triangle of any shape. Bisect each of its angles by construction. If your drawing is correct, the bisectors meet at a point O.

From O draw a perpendicular (with set squares) to a side. With O as centre, and this perpendicular as radius, draw a circle. This circle should *touch* each of the three sides. It is said to be **inscribed** in the triangle.

(Practical Applications. Heights and Distances.)

The following problems are to be solved by measuring diagrams carefully drawn to scale. Since however it is impossible either to draw or to measure with absolute accuracy, it follows that results so obtained can only be **approximate**; that is to say, they will be near enough to the truth to be of practical value, though they cannot be relied upon as strictly accurate. Careful work should usually yield a result within *one* per cent. of that given in the *Answers.*

The direction which we call **vertical** (or *upright*) is that taken by a thread from one end of which a weight hangs freely at rest. Any straight line at right angles to a vertical line is said to be **horizontal** (or *level*).

Ex. 26. How many vertical lines can be drawn through a given point? How many horizontal lines?

In the diagram given below P represents some object whose height or distance is to be found, and O the position of the observer's eye; so that OP is the *line of sight*, that is, the direction in which the object is seen. Let OA be the *horizontal* line passing from the observer's eye directly *under* or *over* the object P.

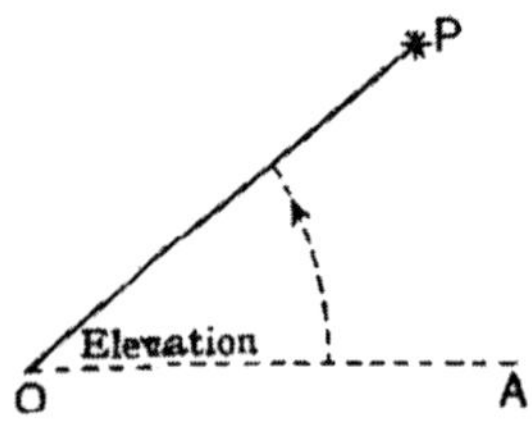

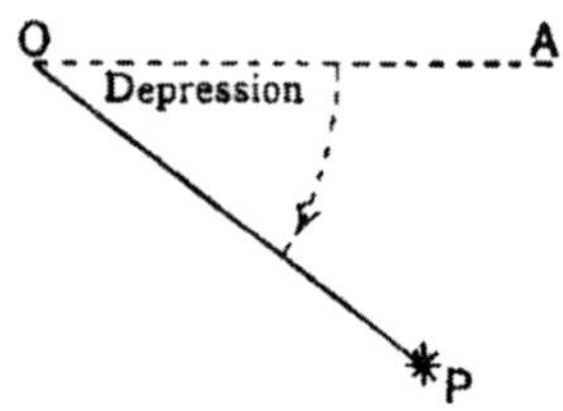

Then the ∠AOP is called the **angle of elevation,** when the object is *above* the horizontal line; and the **angle of depression** when the object is below the horizontal line.

Ex. 27. On my estate there are two farms. One lies S.E. of my house, and 350 yards from it; the other lies S.W. of the house at a distance of 250 yards. How far are the farms apart? (Scale 100 yards to 1 inch.)

Ex. 28. Havre lies due West of Rouen, distant 72 kilometres. Dieppe lies due North of Rouen, distant 56 kilometres. How does Dieppe bear from Havre, and what is the distance between the two places? (Scale 10 km. to 1 cm.)

Ex. 29. A shore battery, whose guns have an effective range of 7000 yards (say 4 miles), fires on an enemy's ship bearing N.W. from the battery and distant $2\frac{1}{2}$ miles. On this the ship steams N.E., 2 miles, then drops anchor, thinking herself out of range. Is she? (Scale 1 mile to 1 inch.)

Ex. 30. A tower is observed from a point on the ground 500 feet distant from its foot, and the angle of elevation of the top is found to be 15°. What is the height of the tower? (Scale 100 feet to 1 inch.)

Ex. 31. A vertical pole, 21 feet high, is found to cast a shadow 35 feet long. How many degrees is the sun above the horizon? (Scale 10 feet to 1 inch.)

Ex. 32. From a point **A** I walk 200 yards due West: I then turn N.E., and walk till I get to a point **C** from which **A** appears due South. Then I return straight to **A**. How far have I walked altogether? (Scale 100 yards to 1 inch.)

Ex. 33. A balloon, held captive by a rope 200 metres long, has drifted in the wind till its angle of elevation, as observed from the place of ascent, is 54°. How high is the balloon above the ground? (Scale 20 metres to 1 cm.)

Ex. 34. From a vessel's fore-top, 80 feet above the sea, a buoy is observed, and the angle of depression found to be 9°. How far is the buoy from the ship? (Scale 100 feet to 1 inch.)

Ex. 35. In surveying an estate I note three cottages **A**, **B**, and **C**. I walk from **A** due East to **B**, the distance being 350 metres, and **C** is on my left hand. The distance from **A** to **C** is 120 metres, and from **B** to **C** 370 metres. In what direction does **C** bear from **A**? (Scale 100 metres to 2 cm.)

Ex. 36. A triangular field is enclosed by two hedges and a ditch. The hedges are each 150 yards long, and they make an angle of 64°. Draw a plan (scale 50 yards to 1 inch), and find the length of the ditch.

Ex. 37. From Dover the bearing of Calais is E. 31° S.; that of Boulogne is E. 63° S.; and the distances of the two French ports from Dover are respectively 23 miles and 31 miles. How far is Boulogne from Calais? (Scale 10 miles to 1 inch.)

Ex. 38. A straight canal runs through my grounds, and is bridged at two places 400 yards apart. The house is 250 yards from each bridge. How far is it from the house to the nearest point on the canal? (Choose a suitable scale for yourself.)

Ex. 39. Two ships A and B drop anchor, 2 cable's lengths apart, B bearing N.W. from A. A signal station ashore bears N.E. from A and due E. from B. How far is each ship from the signal station? [N.B. 1 cable = 200 yards.]

Ex. 40. There are three towns A, B, and C. Of these, B is East of A, and distant 35 miles; while C is North of A, and distant 84 miles. A straight railway connects B and C. How far is A from the nearest point on this railway? (Scale 10 miles to 1 cm.)

Ex. 41. From a certain point on the ground I observe the top of a spire, and find the angle of elevation to be 33°. I advance 80 feet towards the spire, and then find the angle of elevation to be 47°. How high is the spire? (Scale 40 ft. to 1 inch.)

Ex. 42. A man, standing 15 feet away from the base of a monument, finds that the angle of elevation of the summit is 45°; and in making the observation his eye is 5 feet above the level of the ground. Find the height of the monument. (Scale 5 feet to 1 inch.)

Ex. 43. If a man, whose height is 6 feet, stands 12 feet from a certain lamp-post, he finds that his shadow cast by the light is 12 feet in length. How high is the light above the ground?

Ex. 44. From a point on a plain I observe a beacon which stands on the summit of a neighbouring hill, and I find its angle of elevation to be 14°. I walk 700 metres over the plain towards the hill, and then find the angle of elevation to be 31°. How high is the beacon above the level of the plain?

XI. QUADRILATERALS.

Any figure bounded by four straight sides is called a **quadrilateral.**

Before attempting to draw a quadrilateral from given sides and angles be sure to make a rough preliminary freehand sketch, writing in the given dimensions. This will shew you clearly what is given and what is required. In this Section, set squares are to be used for drawing parallels and perpendiculars unless otherwise stated.

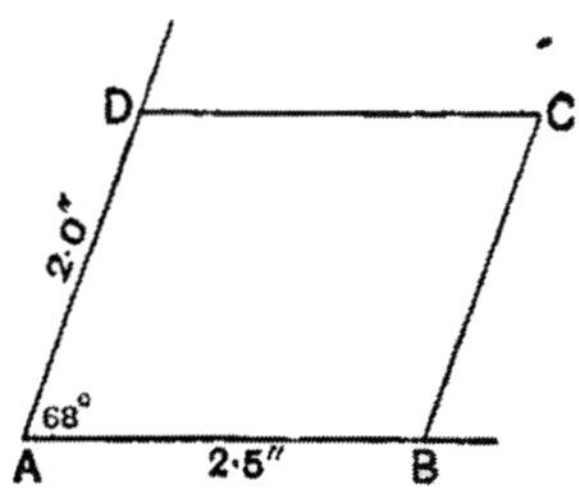

Draw two lines making at A an angle of 68°. Along one arm mark off AB equal to 2·5″; and along the other mark off AD equal to 2·0″.

Through B draw a line parallel to AD.

Through D draw a line parallel to AB, cutting the first parallel at C.

The four-sided figure you have thus drawn is called a **parallelogram.**

A **parallelogram** is a quadrilateral *whose opposite sides are parallel.*

Measure DC and BC, and compare them with AB and AD.

Can you tell from what you have learned of parallels how many degrees there are in the angles ABC, ADC, BCD? Test your answer by measurement.

Ex. 1. Draw a parallelogram ABCD from the following data: The $\angle A = 114°$, $AB = 7·5$ cm., $AD = 5·5$ cm.

Measure DC and BC, and compare them with the given sides.

Write down the number of degrees in each of the angles ABC, ADC, BCD; and test by measurement.

Ex. 2. Draw a parallelogram ABCD in which the $\angle B = 42°$, $AB = 8·2$ cm. and $BC = 6·4$ cm.

Measure and compare (i) the opposite sides, (ii) the opposite angles; and write down the results you get.

Ex. 3. Draw a parallelogram ABCD in which the ∠A is a *right angle*, AB=2·8″, AD=1·7″.

Measure DC and BC, and compare them with the given sides.

What are the other angles of the figure, and why?

Ex. 4. Draw a parallelogram ABCD from the following data:

The ∠A=a *right angle*, and AB=AD=6·5 cm.

Measure DC and BC: do you find all the sides equal?

What are the remaining angles of the figure, and why?

A parallelogram which has a right angle is called a **rectangle.**

A rectangle in which two sides forming a right angle are equal is called a **square.**

Ex. 5. Draw a parallelogram ABCD, in which the ∠A=122°, and AB=AD=7 cm.

Measure DC and BC: do you find the sides all equal?

Write down the number of degrees in each of the angles ABC, BCD, ADC.

A **rhombus** is a parallelogram in which two sides which meet are equal, but it has no right angle.

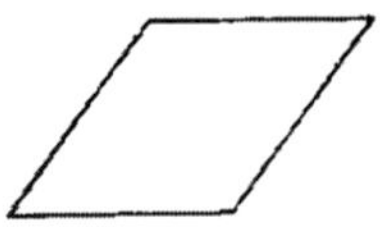

Notice that the *rectangle*, the *square*, and the *rhombus* are all special forms of the parallelogram.

We will now gather together the conclusions that may be drawn from the foregoing exercises.

(i) In a parallelogram, what do you infer about the opposite sides? What about the opposite angles?

(ii) Are all the sides of a square equal? Why?

(iii) Are all the sides of a rhombus equal? Why?

(iv) If one angle of a parallelogram is a right angle, what can you tell about the other angles?

(v) What do you conclude about the angles of a *rectangle*? What about the angles of a *square*?

Each of the straight lines which join opposite vertices of a quadrilateral is called a **diagonal**.

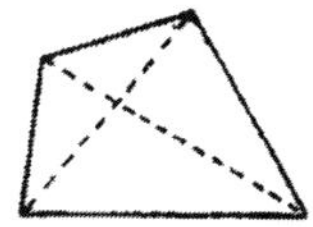

Ex. 6. Draw an *oblique parallelogram* (that is, having no right angle), a *rectangle*, a *square*, and a *rhombus*. Call each figure ABCD. In each case draw the two diagonals, and let them cross at O. Now ascertain by measuring or other experiment to which of these four figures the following statements apply :

(i) *The diagonals bisect one another.*

(ii) *The diagonals cross at right angles.*

(iii) *The diagonals are equal.*

(iv) *Each diagonal divides the figure into two triangles of the same size and shape.* (Make a tracing of the △ ABC, and see if it can be exactly fitted over the △ ADC.)

(v) *The figure is symmetrical about a diagonal.* (That is, if the figure is folded about a diagonal, the two parts coincide.)

Ex. 7. About which of the four figures of Ex. 6 can a circle be circumscribed having its centre at O, and OA as radius?

Ex. 8. Using your protractor and set squares, draw a rhombus having each side 6·5 cm. in length, and one angle equal to 82°. Enter into your figure (without measurement) the values of the other angles.

Ex. 9. On a side of 2·5″ construct a square with ruler and compasses only. Measure each diagonal to the nearest tenth of an inch.

Ex. 10. Draw a line AC, 3″ long. With ruler and compasses only construct a square having AC as diagonal; and measure its sides.

[First step of construction : Bisect AC at right angles.]

Ex. 11. Construct a rhombus whose diagonals are 8 cm. and 6 cm. (using ruler and compasses). Measure each side.

Ex. 12. Draw a parallelogram ABCD, in which the sides AB, AD are 6·5 cm. and 5·5 cm., and the diagonal BD is 9 cm.

[Construct the △ ABD (Problem 11); then complete the parallelogram with set squares.]

Problem 14.

To construct a quadrilateral **ABCD**, *having the angle at* **A** *equal to a given angle* **P**, *and the sides of given lengths.* (For instance: **AB** = 8 cm., **BC** = 5 cm., **CD** = 6 cm., **DA** = 7 cm.)

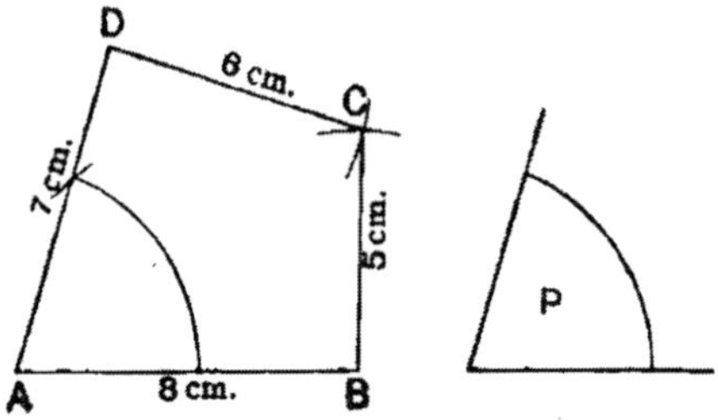

Construction. Construct an angle at **A** equal to the given angle **P**; and from its arms cut off **AB** equal to 8 cm., and **AD** equal to 7 cm.

With centre **D**, and radius 6 cm., draw an arc.

With centre **B**, and radius 5 cm., cut the first arc at **C**.

Join **BC**, **DC**.

Then **ABCD** is the required quadrilateral.

Note. If the ∠**A** is given in *degrees*, it must be made with the protractor.

Ex. 13. Draw quadrilaterals from the rough plans given below; the dimensions are to be in centimetres.

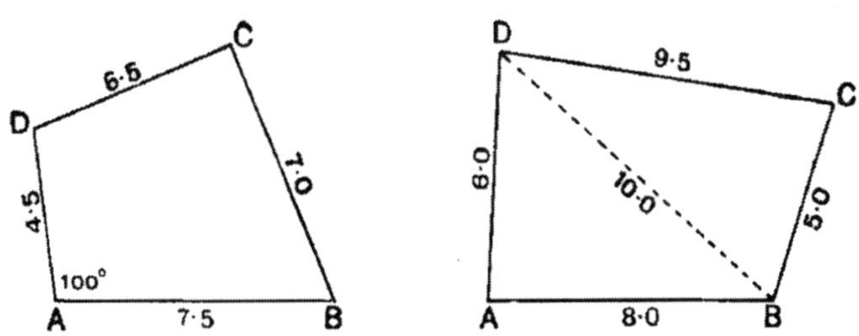

[In the right-hand figure first construct the △**ABD** (Problem 11), then proceed as above.]

Ex. 14. In a quadrilateral ABCD,

$$AB = 3{\cdot}5'', \quad BC = 3{\cdot}0'', \quad CD = 2{\cdot}5'', \quad DA = 2{\cdot}0''.$$

Shew that the *shape* of the figure is not fixed by these data.

Draw the quadrilateral from the above dimensions, when

(i) $A = 60°$; (ii) $A = 90°$.

How many things must be given in order to fix the size and shape of a quadrilateral?

Ex. 15. In surveying a quadrilateral field ABCD, I go from A to B due East, and find that AB = 50 metres; from B to C North East, and BC = 60 metres; from C to D due West, and CD = 135 metres.

Plot the field, (scale 10 metres to 1 cm.). Measure DA on your plan: what is the real length of this side? How does D bear from A? Shew by any test you like that the sides AB, CD are parallel.

A quadrilateral that has *one* pair of parallel sides is called a **trapezium**.

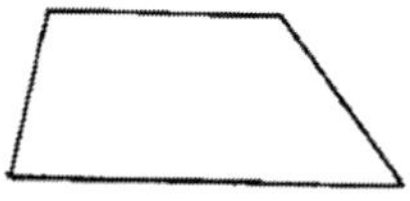

Ex. 16. Draw a parallelogram whose diagonals are 8 cm. and 6 cm. in length, and intersect one another at an angle of 54°. Find by measurement the length of the perimeter.

Ex. 17. I want a plan of a quadrilateral field ABCD, and I have with me no means of measuring *angles*.

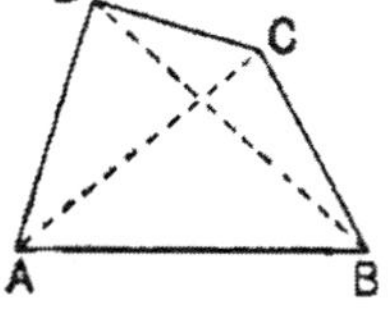

I therefore measure the following lengths:

AB = 350 yards, AC = 300 yards, BC = 200 yards;
AD = 230 yards, BD = 350 yards.

Plot the field from these dimensions, and measure the side CD.

[First construct the △ ABC (Problem 11), scale 100 yards to 1 in.; then construct the △ ABD. Finally join CD.]

Ex. 18. Draw a good sized quadrilateral **ABCD** of any shape; and make an exact copy of it by each of the following constructions:

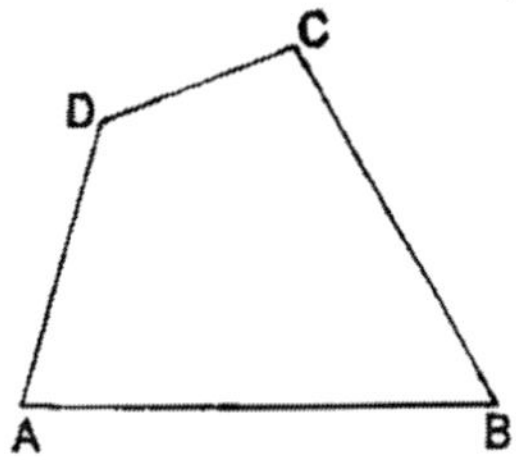

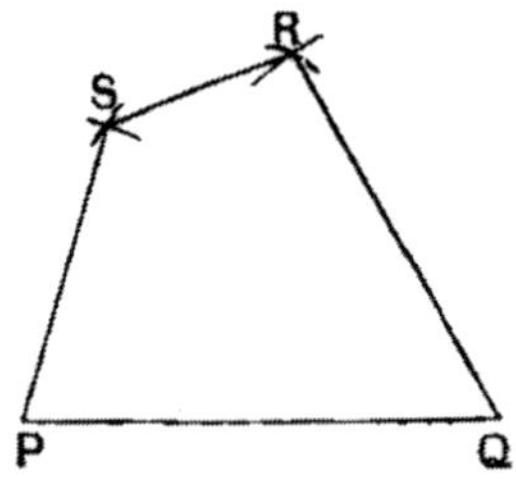

Draw **PQ** equal to **AB**.

(i) Make the ∠**P** equal to the ∠**A** (with your protractor, if this is allowed; otherwise by construction). Cut off **PS** equal to **AD**.

Make the ∠**PQR** equal to the ∠**B**; and cut off **QR** equal to **BC**.

Join **SR**.

(ii) With centre **P**, and radius equal to **AD**, draw an arc.

With centre **Q**, and radius equal to **BD**, cut the first arc at **S**.

With centre **P**, and radius equal to **AC**, draw an arc.

With centre **Q**, and radius equal to **BC**, cut the last arc at **R**.

Join **SR**.

NOTE. A figure of *five*, or more, sides may be reproduced by similar constructions.

Ex. 19. Draw the patterns shewn below:

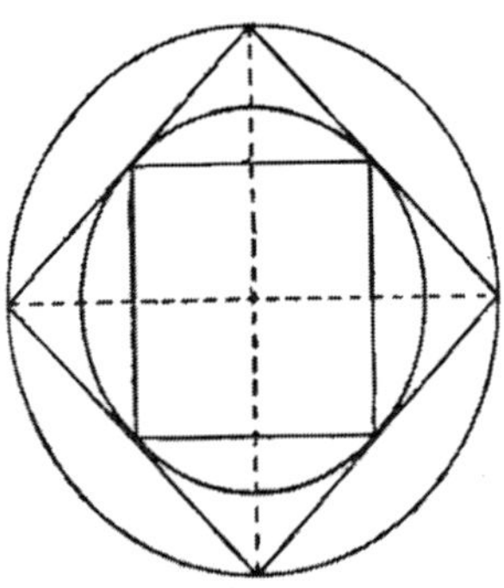

Fig. 1.

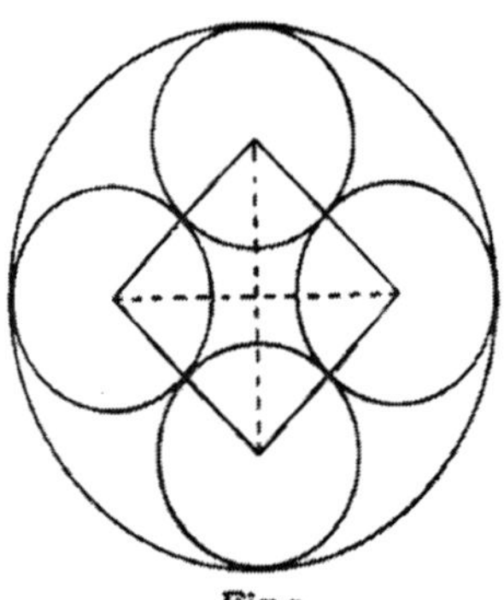

Fig. 2.

Fig. 1. First draw the larger square from diagonals of 3″.

Fig. 2. First draw the square from diagonals of 2″; then the small circles; finally the outside circle.

XII. AREAS.

In the squared paper used in this Section the horizontal lines are *one-tenth* of an inch apart, and the perpendicular lines are also *one-tenth* of an inch apart; so that the whole surface of the paper is divided into little squares, each on a side of one-tenth of an inch.

The figure **ABCD** is a rectangle whose length **AB** is 1·5", and whose breadth **AD** is 0·8"; so that the length and breadth contain respectively 15 and 8 *tenths of an inch.*

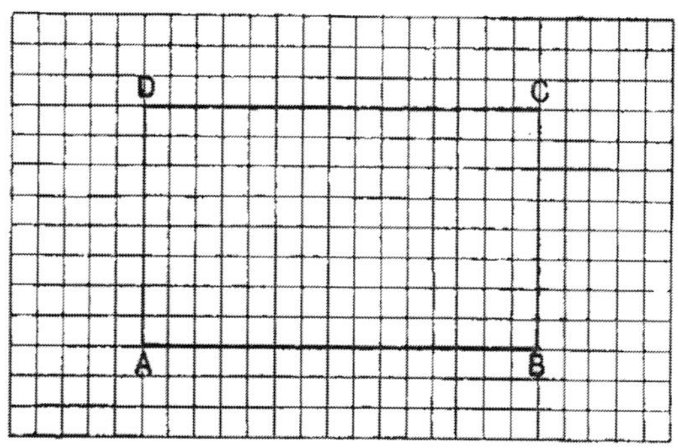

Now let us reckon the number of squares that fall within this rectangle.

These squares lie in *rows* parallel to **AB**. How many squares are there in each row? How many rows are there? How many squares then are there altogether in the rectangle?

Again the squares stand in *columns* parallel to **AD**. How many squares are there in each column? How many columns? How many squares altogether?

The total number of squares within the rectangle gives you an idea of the **area**, that is to say, the *amount of space* enclosed within its boundaries.

Ex. 1. Draw on squared paper a rectangle whose length is 2·0″, or 20 *tenths* of an inch, and whose breadth is 0·9″, or 9 *tenths.*

Count the number of squares in each row, and the number of rows. How many squares are there in the rectangle?

Check your answer by counting the number of squares in each column, and the number of columns.

Ex. 2. Draw the following rectangles on squared paper, and find their areas (measured in squares on one-tenth of an inch):

(i) Length = 2·0″, breadth = 1·0″;

(ii) Length = 1·5″, breadth = 1·2″;

(iii) Length = 2·5″, breadth = 0·8″;

(iv) Length = 1·6″, breadth = 0·5″.

State a rule by which you can find the number of squares in each rectangle without counting them all.

Ex. 3. Draw on squared paper a rectangle of length 1·2″, making the breadth such that the rectangle will contain 84 ruled squares.

Ex. 4. Consider the figures given on the opposite page. Which do you think contains the greatest area? Which the least?

Now count the squares in each figure, and see if your guess is correct.

[In several of these figures the outlines run *through* some of the squares. In such cases

portions of a square which seem to be one-half should be counted as *half-squares*;

portions which seem greater than one-half should be counted as *whole squares*;

portions which seem less than one-half should be omitted.

This is, of course, a somewhat rough and ready way of counting, and results so obtained cannot be expected to be quite correct; but they will be near enough for our present purpose.]

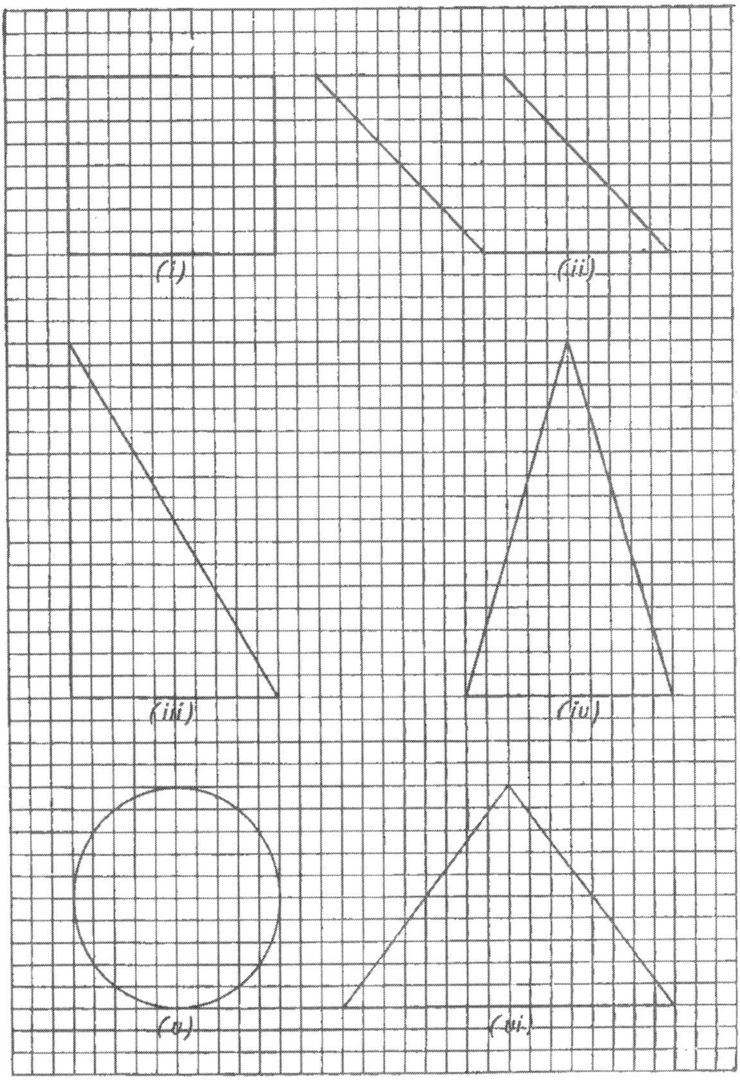

From these examples you see that figures may differ completely in shape, and yet *contain the same amount of space* within their boundaries.

The amount of space contained in a square drawn on a side one inch in length is called a **square inch.**

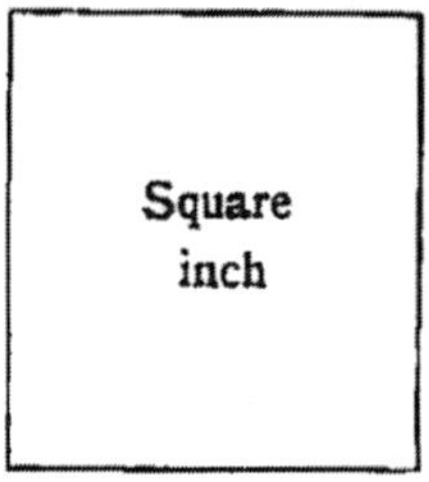

Again a **square centimetre** is the area of a square drawn on a side of one centimetre.

The terms **square yard, square foot, square metre** have similar meanings.

We measure the area of a figure by noting how many square inches, or square centimetres, or other such units of area it contains.

NOTE. You will clearly understand that a figure containing an area of 1 *square inch* is not itself necessarily square: it may be triangular, or circular, or of any other shape, provided that its boundaries enclose exactly as much space as that contained within a square on a side of 1 inch.

(Areas of squares and rectangles.)

Ex. 5. Draw on squared paper a square on a side of 1 inch. How many squares does it contain, each on a side of *one-tenth* of an inch?

Ex. 6. Draw two straight lines, one double the length of the other; and on each draw a square. How many times does the greater square contain the less? Draw lines in the greater square to illustrate your answer.

Ex. 7. Draw on squared paper a rectangle 1·5″ long by 1·0″ wide.

(i) If you treble the length, without altering the width, how many times do you multiply the area?

(ii) If you treble both length and breadth, how many times do you multiply the area?

(iii) If you treble the length, and double the breadth, how many times do you multiply the area?

In each case draw a figure to illustrate your answer.

Ex. 8. Draw a line AB, 3″ long. Suppose each inch to stand for 1 foot, so that the whole line represents 1 yard.

Draw a square on AB: then this square represents 1 *square yard*. In the corner of this figure draw a square to represent 1 square foot.

Now shew why 1 square yard = 9 square feet.

Ex. 9. A passage is 20 feet long by 10 feet wide. Draw a plan of the floor on squared paper (scale 10 feet to 1 inch).

How is a square foot represented on your plan? Find the area of the floor in square feet.

Ex. 10. A court-yard is 25 yards long by 15 yards wide. Draw a plan on squared paper (scale 10 yards to 1 inch).

What area is represented by one of the ruled squares of your paper? Find the area of the court-yard.

Ex. 11. Find the area of the rectangles of which the length and breadth are given below.

The areas are to be got by calculation; but it will be a useful exercise to draw a plan on squared paper in each case. Choose your own scale.

(i)	Length = 18 in.,	breadth = 10 in.;	Ans. in sq. in.
(ii)	Length = 25 ft.,	breadth = 16 ft.;	Ans. in sq. ft.
(iii)	Length = 45 metres,	breadth = 22 metres;	Ans. in sq. m.
(iv)	Length = 2 ft. 1 in.,	breadth = 8 in.;	Ans. in sq. in.
(v)	Length = 5 ft.,	breadth = 48 in.;	Ans. in sq. ft.

Ex. 12. The area of a rectangle is 6 square inches, and its length is 3 inches. What is its breadth? Draw the rectangle.

Ex. 13. Draw a rectangle 5 cm. long, and of sufficient breadth to give the figure an area of 20 sq. cm.

Ex. 14. What is the breadth of a rectangle, if its area is 4 sq. in., and its length is $2\frac{1}{2}''$? Draw the rectangle on squared paper, and thus verify your work.

Ex. 15. If in a plan 1 inch represents 8 feet, what does 1 *square inch* represent?

Ex. 16. In the plan of a quadrangle 1 inch stands for 10 feet: what is represented by 1 *square inch*?

If the length of the plan is 5″, and the breadth is 4″, what is the area of the quadrangle?

Ex. 17. Find the area of a rectangular pavement, of which a plan, scale 5 feet to 1 in., measures 8″ long by 6″ wide.

Ex. 18. In a certain map 1″ represents 5 miles: what area is represented by a rectangle 2·5″ long by 2·0″ wide?

Ex. 19. If 100 yards of railing are required to fence in a square paddock, what is its area?

Ex. 20. The length of a rectangular field is 50 yards: the total distance round it is 180 yards. What is the breadth? Find the area.

(*Area of a parallelogram.*)

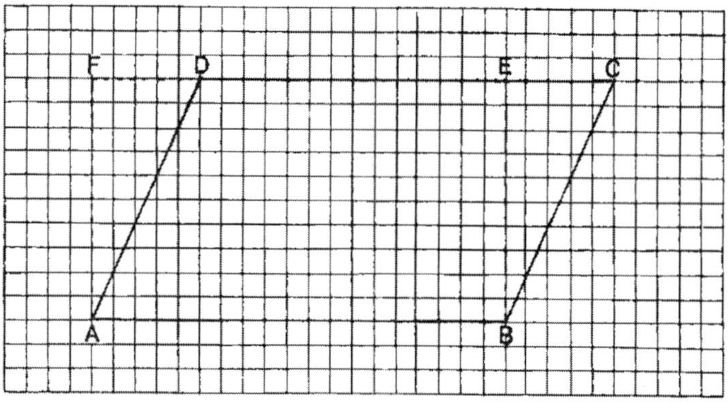

We wish to ascertain the area of the parallelogram ABCD, and in particular to compare it with that of the rectangle ABEF on the same base AB and of the same height BE.

(i) Count the number of ruled squares in the parallelogram, as explained on p. 70.

Then measure the length AB and the height BE in tenths of an inch. Multiply together the number of tenths in the length and height. The product gives you the number of ruled squares in the rectangle ABEF.

Do you get the same, or nearly the same, result for the parallelogram and rectangle? (Remember that your system of counting in the first case is not likely to give a quite correct answer.)

(ii) Make a careful copy of the parallelogram ABCD, and cut it out from your paper. Next rule the line BE; and, cutting along it, remove the triangle BEC.

Now place the triangle BEC on the other side of the remaining figure ABED, so that BC fits along AD.

You see that by thus changing the position of the triangle BEC you have converted the parallelogram into a rectangle.

We conclude that *in this case* the area of the parallelogram is equal to that of the rectangle on the same base and of the same height.

This we may express by saying, that

the area of a parallelogram = base × height.

[For a general and formal proof of this, see *School Geometry*, pp. 104, 105.]

Ex. 21. Draw on squared paper any oblique parallelogram whose base measures 15, and height 8 tenths of an inch.

Draw a rectangle of equal area; and test your work by counting the number of ruled squares in each figure.

Ex. 22. Draw any oblique parallelogram having a base of 3·0″, and height of 2·0″; then draw a rectangle of equal area.

How many such parallelograms could be drawn? How can we tell that they must all have the same area?

Ex. 23. Draw a parallelogram ABCD, in which the length = 8 cm., the height = 5 cm., and the ∠A = 45° (with protractor).

Cut your figure out, and by dissection convert it into a rectangle of the same base and height.

Ex. 24. Rule on your squared paper a rectangle of length 2·5″ and breadth 2·0″.

On the same base draw a parallelogram having the same height as the rectangle, and one angle equal to 60°. (Use your protractor.)

Find the area of each figure.

Ex. 25. Given a square on a side of 6 cm., draw a parallelogram of equal area on the same base, having an angle of 75°.

What is the area of each figure?

Ex. 26. On a base of 2·0″ draw a rhombus having an angle of 50° ; and on the same base draw a rectangle of equal area.

Measure the breadth of the rectangle, and hence calculate the area of each figure.

Ex. 27. Give a construction for drawing a parallelogram ABCD, having two adjacent sides AB, AD equal to 7 cm. and 6 cm. respectively, and having a height of 4 cm. Find its area.

Ex. 28. Draw a parallelogram ABCD, in which AB=8 cm. AD=6 cm., and the ∠A=72°.

Measure the height of the figure, and hence calculate its area.

(*Area of a triangle.*)

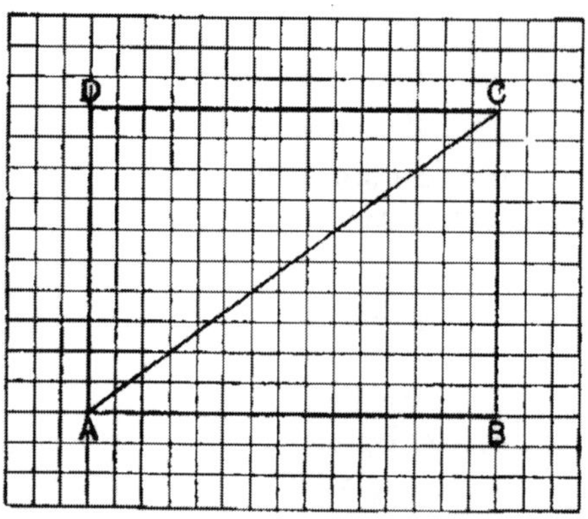

In the rectangle ABCD we have drawn the diagonal AC, thus dividing the figure into two *right-angled* triangles.

We have already found (Ex. 6, p. 65) that these triangles are equal in all respects, so that the area of each is half that of the triangle.

This being so, we can find the number of ruled squares in the triangle ABC by calculating the number in the rectangle ABCD, and then taking half the result. Test this by counting the squares in the triangle ABC.

In the next figure the triangle ABC is not right-angled; but BCDE is still the rectangle on the same base BC and of the same height AF.

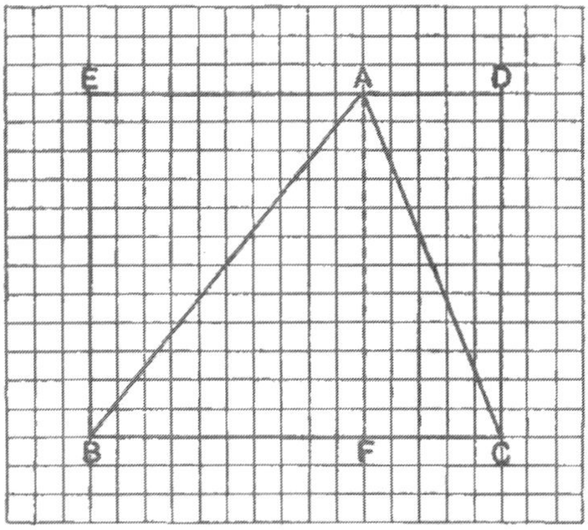

Now the △ AFB is half the rectangle AFBE; why?
And the △ AFC is half the rectangle AFCD; why?
Hence if we add these areas together, we see that the △ ABC is half the rectangle BCDE.

That is to say,

area of △ABC = $\frac{1}{2}$ (*base* BC × *height* AF).

Ex. 29. Draw on squared paper a right-angled triangle ABC, making the right angle at B; and make BC = 3·0″, and BA = 2·0″.

Now complete the rectangle ABCD. What is its area? What is the area of the triangle ABC?

Ex. 30. On your squared paper rule a square on a side of 2·0″. Draw a right-angled triangle of half the area.

Test your work by counting the ruled squares.

Ex. 31. On a base of 8·0 cm. draw a triangle of height 5·0 cm.; then draw a rectangle of double the area.

What is the area of the triangle in square centimetres?

Ex. 32. Draw on squared paper *any* triangle having a base of 2·5″ and a height of 1·6″; and rule a rectangle of double the area.

How many such triangles could be drawn? How can we tell that they are all of equal area?

Ex. 33. Given a rectangle measuring 2·8″ by 1·5″ (rule this on squared paper): on the longer side as base draw an isosceles triangle of half the area.

Ex. 34. Rule on squared paper a rectangle ABCD, in which AB=3·0″, and AD=1·8″.

On the base AB draw two triangles APB, AQB, each having half the area of the rectangle :

in (i) AP is to be 2·6″ in length (by compass construction);

in (ii) the ∠QAB is to be 42° (use protractor).

Ex. 35. One side of a triangular field measures 120 yards, and the shortest distance between the opposite corner and this side is 80 yards. Find the area of the field in square yards.

Ex. 36. There are two plots of ground, one triangular and the other square.

The largest side of the triangular plot is 96 metres, and the perpendicular on it from the opposite corner is 27 metres.

Each side of the square plot measures 36 metres.

Which plot has the larger area?

Ex. 37. Draw an equilateral triangle ABC on a base of 2·0″.

Drop a perpendicular AD from the vertex A on the base BC. Measure AD as accurately as you can.

Now find approximately the area of the triangle in square inches.

Ex. 38. Construct a triangle ABC, having given $a=8$ cm., $b=7$ cm., $c=6$ cm.

Draw and measure the perpendicular from A on BC, and hence calculate the approximate area of the triangle.

Ex. 39. Draw a triangle ABC from the following data :

$$a=7·2 \text{ cm.}, \quad B=68°, \quad C=54°.$$

Draw and measure the perpendicular from A on BC, and hence reckon the approximate area of the triangle.

Ex. 40. (i) Draw an oblique parallelogram, a rectangle, and a triangle, all on the same base and of the same height.

How can you shew from this figure that the area of the triangle is half that of the parallelogram?

(ii) Draw any triangle ABC (not isosceles) on a given base BC. Bisect the base at X, and join AX.

How can you tell that the two triangles ABX, ACX, though of different shape, have the same area?

XIII. MISCELLANEOUS CONSTRUCTIONS. CIRCLES. REGULAR POLYGONS.

PROBLEM 15.

Given the circumference of a circle, to find its centre.

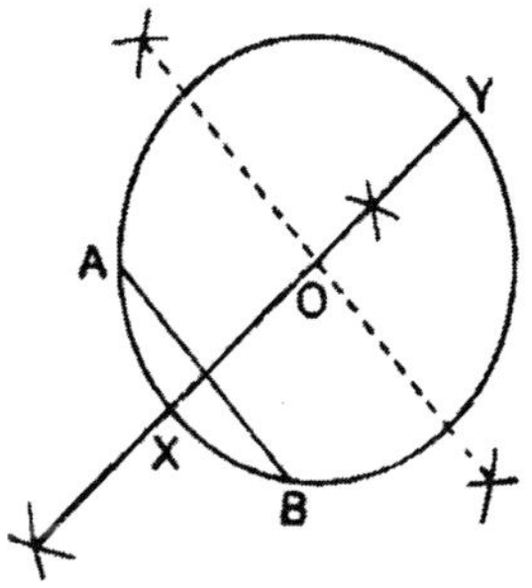

Construction. Draw any chord AB.

Bisect AB at right angles (Prob. 7, p. 41) by a line which cuts the circumference at X and Y.

Bisect XY at O. Then O is the centre.

(Verification.)

In Ex. 5, p. 42 you found several points whose distances from two given points A and B were equal; and on comparison, it appeared that all such points lay on a certain line: what line?

Now the centre is a point whose distances from A and B are equal; so the centre lies on that same line. Thus we have been able to draw a diameter XY; and by bisecting this, to find the centre.

(*The Angle in a Semi-circle.*)

Draw a good sized semi-circle, say of radius 6 cm., and call its diameter AB. On the semi-circumference take three or four points P, Q, R, ... ; and join each of them to A and B.

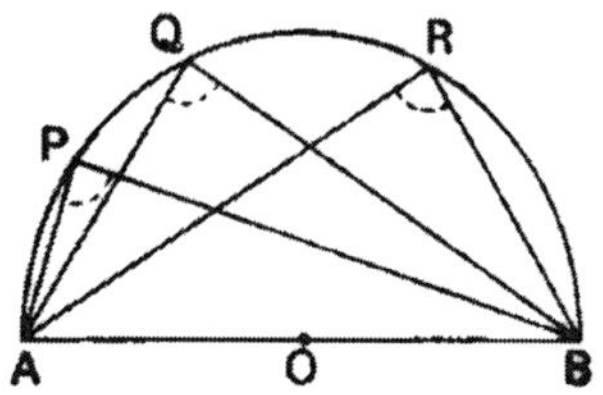

Now measure the angles APB, AQB, ARB ; and enter the results in your figure.

Repeat this experiment with another semi-circle of any size you please, and again record your results.

You have now, no doubt, found in the instances you have examined, that if you join a point on the semi-circumference to the ends of the diameter, the angle so formed is *a right-angle.*

This result we express by saying that *the angle in a semi-circle is a right angle.*

Several important constructions follow from this property of a semi-circle.

Ex. 1. *Draw a straight line perpendicular to a given straight line* AB *from a given point* X *outside it,* X BEING NEARLY OPPOSITE ONE END OF AB.

[In this case the construction of Problem 10, p. 46 is inconvenient, and in its place we may use the following.]

Construction. Take any point D in AB. Join DX ; and bisect DX at C.

On DX draw a semi-circle to cut AB at O.

Join XO, and explain why it is perpendicular to AB.

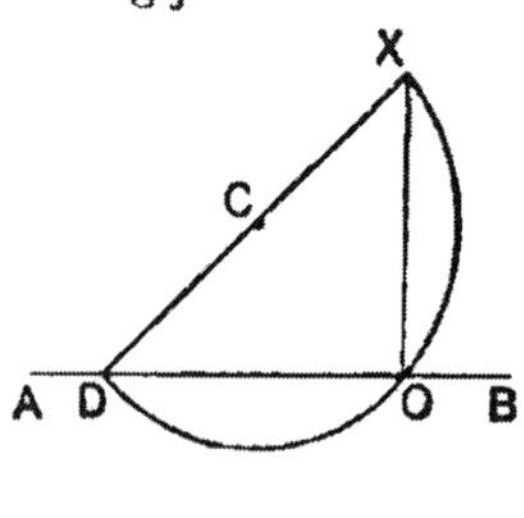

(Tangents to a Circle.)

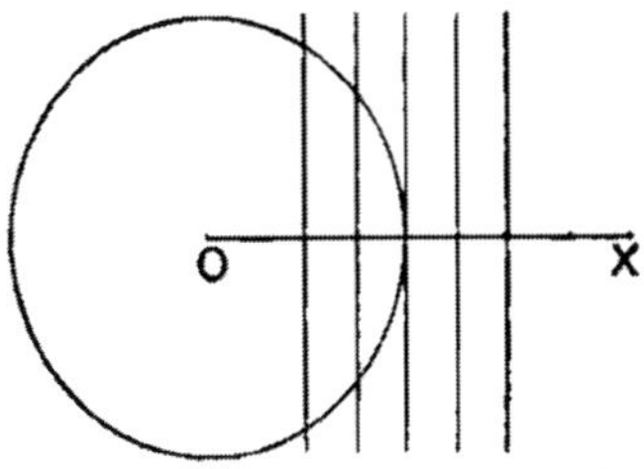

With any point O as centre, and a radius of 4 cm., draw a circle. Draw a radius, and produce it to X.

In OX take points at distances of 2 cm., 3 cm., 4 cm., 5 cm., and 6 cm. from O.

Through these points draw lines with your set squares perpendicular to OX.

Notice if, and how, these perpendiculars meet the circumference.

If the distance of the perpendicular from the centre is *less than the radius*, in how many points does the perpendicular meet the circle? If greater, in how many points? If equal to the radius, in how many points?

From this and similar experiments you may learn that a line drawn perpendicular to a radius through its extremity meets the circumference at *one point only*. Such a line is said to *touch* the circle at that point, and is called a **tangent**.

Observe that only *one* tangent can be drawn to a circle at a given point on its circumference. Why so?

Ex. 2. In a circle of radius 1·8″ draw a diameter AB. Then with your set squares draw tangents at A and B, and shew that these are parallel.

Ex. 3. Draw a straight line AB and take a point X in it. If a circle touches AB at X, on what line must its centre lie?

Draw two circles of radius 3·0 cm. to touch AB on opposite sides at the point X.

Ex. 4. Draw two concentric circles of radii 6·0 cm. and 6·5 cm. Draw a chord of the larger circle to touch the smaller, and measure its length.

PROBLEM 16.

To draw a pair of tangents to a circle from a given point T *outside it.*

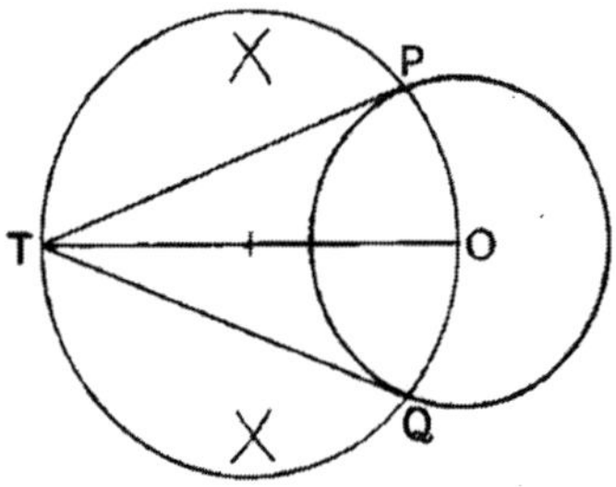

Construction. Join T to O the centre of the given circle, and bisect TO.

On TO as diameter draw a circle cutting the given circle at P and Q.

Draw TP and TQ, which are the required tangents.

(*Verification.*)

Draw the radius OP. Then if the ∠OPT is a right angle, PT is a tangent. Now the ∠OPT is an angle in a semi-circle, and therefore a right angle.

Ex. 5. Draw a circle of radius 1·5″, and from a point 2·5″ distant from the centre draw a pair of tangents to the circle.

Measure the lengths of the tangents, and note that they are equal.

Ex. 6. Draw a circle of radius 4·0 cm. Take any two points P and Q, each at a distance of 10·4 cm. from the centre. From P and Q draw pairs of tangents to the circle.

Shew by measurement that all four tangents are equal.

Ex. 7. Take two points A and B, 3 cm. apart. With A and B as centres, and a radius of 3 cm., draw circles. Produce AB both ways to cut the first circle at X and the other at Y.

From X draw a pair of tangents to the circle whose centre is B.

From Y draw a pair of tangents to the circle whose centre is A.

What sort of quadrilateral is the figure so formed?

A figure bounded by more than four sides is called a **polygon**. It is said to be **regular** if all its sides are equal, and all its angles are equal.

The most important regular polygons are these:

A **pentagon**, which has *five* sides; a **hexagon**, *six* sides; an **octagon**, „ „ *eight* sides; a **decagon**, *ten* sides.

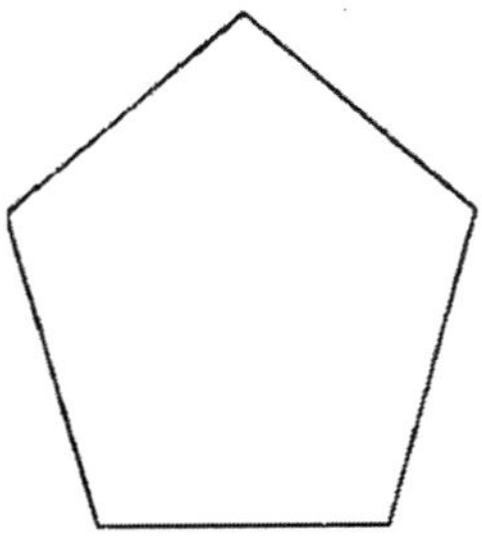

Pentagon

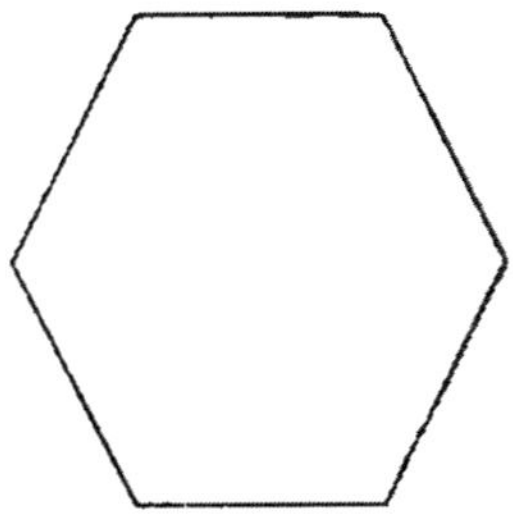

Hexagon

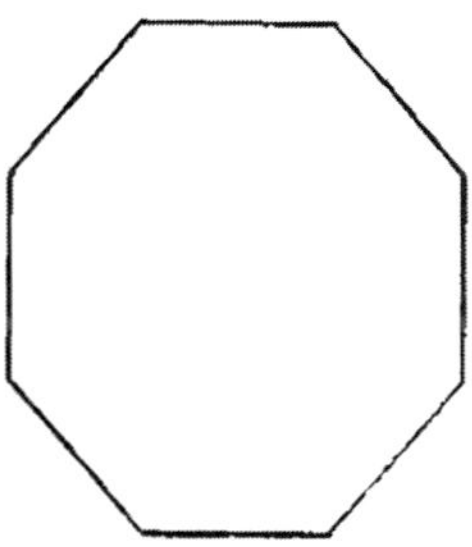

Octagon

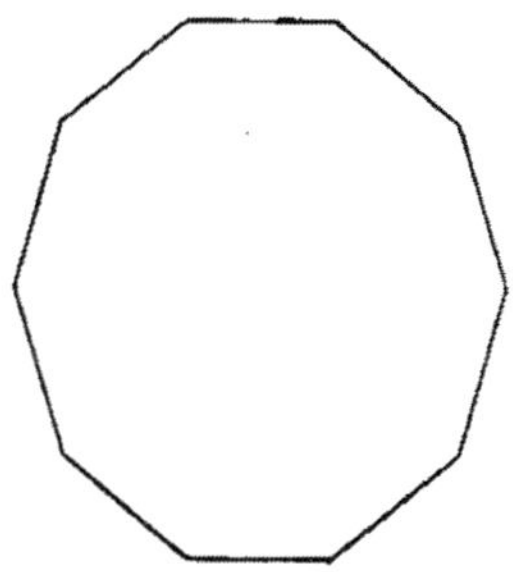

Decagon

Ex. 8. What name has been given to a regular figure of *three* sides? What name to a regular figure of *four* sides?

Let us consider a regular pentagon ABCDE inscribed in a circle.

Join the centre O to each vertex.

How many degrees are there in each of the angles AOB, BOC, COD, DOE, EOA?

The angle AOB is called the **central angle** of the polygon.

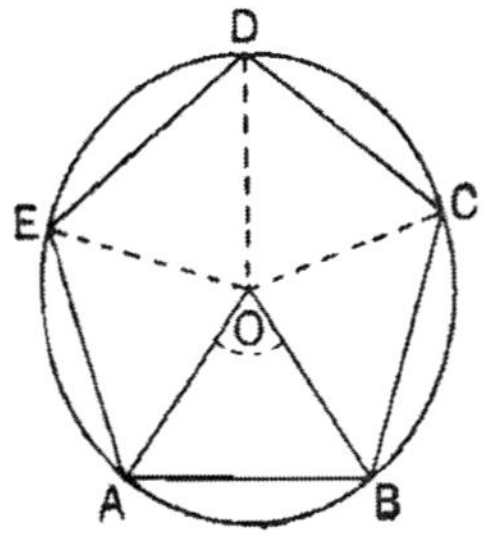

Ex. 9. How many degrees are there in the central angle of (i) an equilateral triangle, (ii) a square, (iii) a regular hexagon, (iv) a regular octagon, (v) a regular decagon?

Which of these angles have you learnt to construct with ruler and compasses only?

It is now evident that to inscribe in a given circle a regular polygon of a given number of sides, we must first draw its *central angle* AOB. This fixes the length of the side, or chord, AB, which may then be stepped off round the circumference the required number of times.

Ex. 10. Draw a circle of radius 4·5 cm., and inscribe in it an equilateral triangle (with ruler and compasses).

Ex. 11. In a circle of radius 4·5 cm. inscribe a square (with ruler and compasses).

Ex. 12. Inscribe a regular pentagon in a circle of diameter 3·6″ (with protractor). Measure any two of its angles.

Ex. 13. Inscribe a regular hexagon in a circle of radius 1·6″ (with ruler and compasses). Measure any two of its angles.

Join each vertex to the centre, and shew by measurements or reasoning that the hexagon consists of *six equilateral triangles.*

Ex. 14. In a circle of diameter 8 cm. inscribe a regular octagon, using your protractor.

Repeat this exercise, using ruler and compasses only.

Ex. 15. Draw a square on a side of 7·0 cm. (with protractor); and find its central point with your ruler.

Draw a circle to pass through all the vertices of the square.

Draw a second circle within the square to touch each of its sides.

Ex. 16. How would you find the central point of an equilateral triangle with ruler and compasses?

Draw an equilateral triangle on a side of 3·0″, and circumscribe a circle about it.

Ex. 17. Draw a circle of radius 1·5″: then draw a square about it, so that each side touches the circle. What is the length of each side?

Ex. 18. *On a side* **AB**, 4 *cm. in length, draw a regular* **hexagon.**

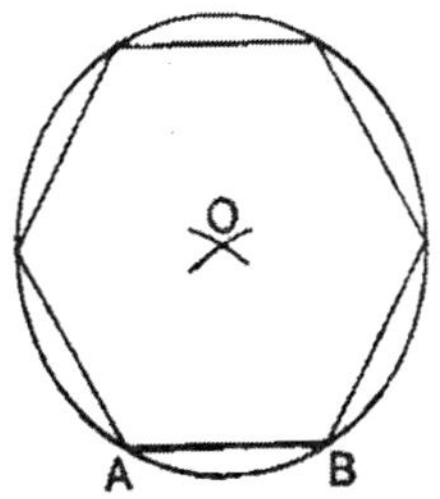

We have seen (Ex. 13) that a regular hexagon is built up of six equilateral triangles, and that its central angle is 60°. This suggests the following construction :

Construction. Find the vertex **O** of an equilateral triangle **AOB**, standing on the base **AB.**

With centre **O** and radius **OA** draw a circle. Then step off chords, each equal to **AB**, round the circumference.

Ex. 19. *On a side* **AB**, 3 *cm. in length, draw a regular* **octagon.**

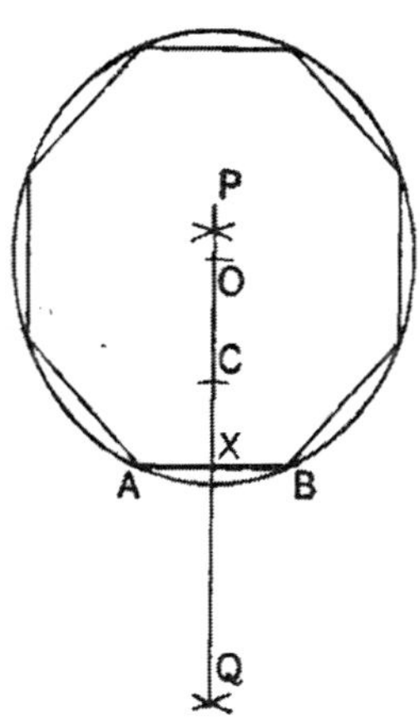

The central angle of a regular octagon is 45° : on this the following construction is based.

Construction. Bisect **AB** at right angles by the line **PQ**, cutting **AB** at **X.**

From **XP** cut off **XC** equal to **XA.**

From **CP** cut off **CO** equal to **CA.**

With centre **O**, and radius **OA**, draw a circle : then step off chords each equal to **AB** round the circumference.

(Verification.)

Join **OA, OB.** We want to see why the construction makes the ∠**AOB** equal to 45°. Join **AC.**

How many degrees are there in the ∠**ACX**, and why ?

How many degrees in the ∠**COA**, and why ?

Now deduce the number of degrees in the ∠**AOB.**

Ex. 20. Shew how to cut off the corners of a square so as to obtain from it a regular octagon.

XIV. THE FORM OF SOME SOLID FIGURES.

(*Rectangular Blocks.*)

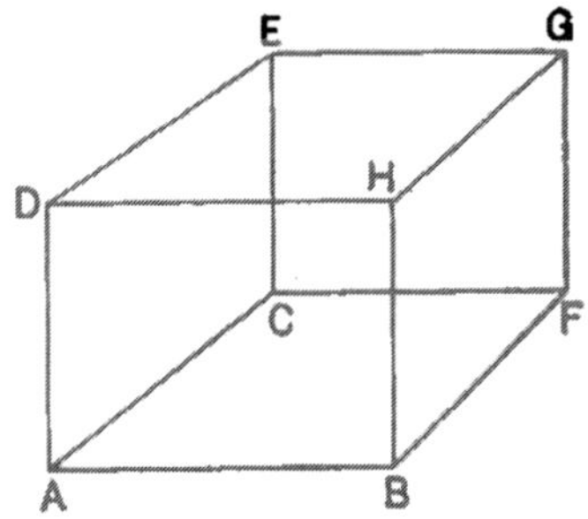

The solid whose shape you are probably most familiar with is that represented by a brick or slab of hewn stone. This solid is called a **rectangular block** or **cuboid.** Let us examine its form more closely.

How many *faces* has it? How many *edges*? How many *corners*, or *vertices*?

The faces are quadrilaterals: of what shape?

Compare two opposite faces. Are they equal? Are they parallel?

We may now sum up our observations thus:

A cuboid has *six* faces; opposite faces being *equal rectangles in parallel planes.* It has *twelve* edges, which fall into three groups, corresponding to the *length*, the *breadth*, and the *height* of the block. The four edges in each group are equal and parallel, and perpendicular to the two faces which they cut.

The length, breadth, and height of a rectangular block are called its **three dimensions.**

Ex. 1. If two dimensions of a rectangular block are equal, say, the breadth AC and the height AD, two faces take a particular shape. Which faces? What shape?

Ex. 2. If the length, breadth, and height of a rectangular block are all equal, what shapes do the faces take?

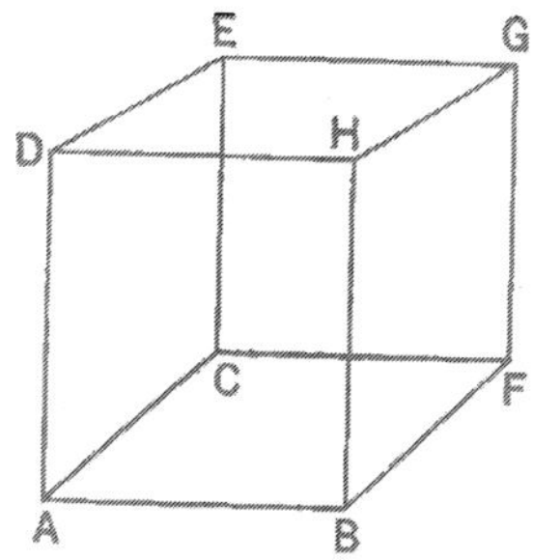

A rectangular block whose length, breadth, and height are all equal is called a **cube**. Its surface consists of six equal squares.

We will now see how models of these solids may be constructed, beginning with the cube, as being the simpler figure.

Suppose the surface of the cube to be cut along the upright edges, and also along the edge HG; and suppose the faces to be unfolded and flattened out on the plane of the base. The surface would then be represented by a figure consisting of six squares arranged as below.

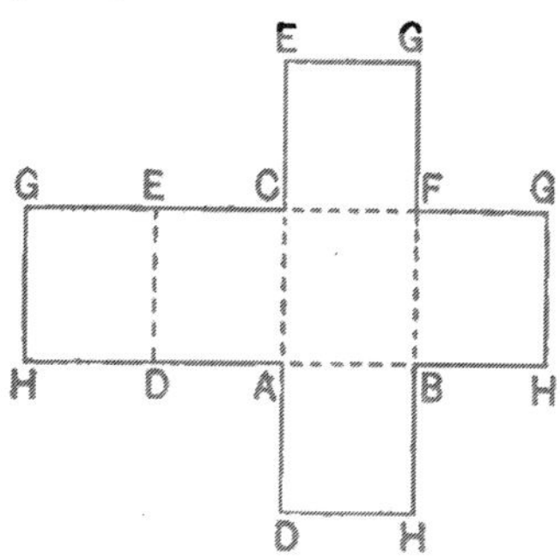

This figure is called **the net** of the cube: it is here drawn on half the scale of the cube shewn in outline above.

To make a model of a cube, draw its net on cardboard. Cut out the net along the outside lines, and cut partly through along the dotted lines. Fold the faces over till the edges come together; then fix the edges in position by strips of gummed paper.

Ex. 3. Make a model of a cube each of whose edges is 6·0 cm.

Ex. 4. Make a model of a rectangular block, whose length is 4″, breadth 3″, height 2″.

First draw the net which will consist of six rectangles arranged as below, and having the dimensions marked in the diagram.

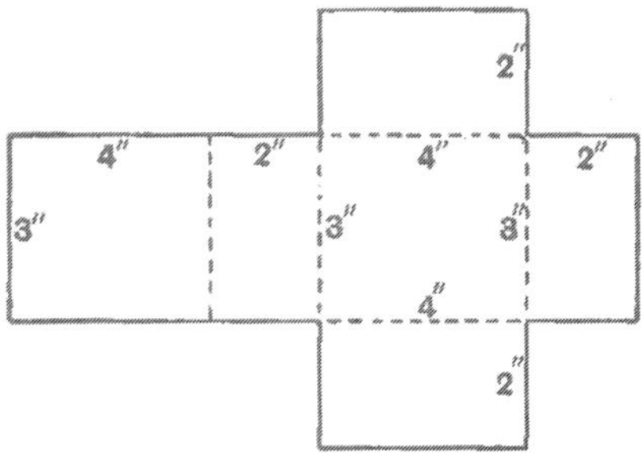

Now cut the net out, fold the faces along the dotted lines, and secure the edges with gummed paper, as already explained.

(*Prisms.*)

Let us now consider a solid whose side-faces (as in a rectangular block) are rectangles, but whose *ends* (*i.e.* base and top), though equal and parallel, are not necessarily *rectangles.* Such a solid is called a **prism**.

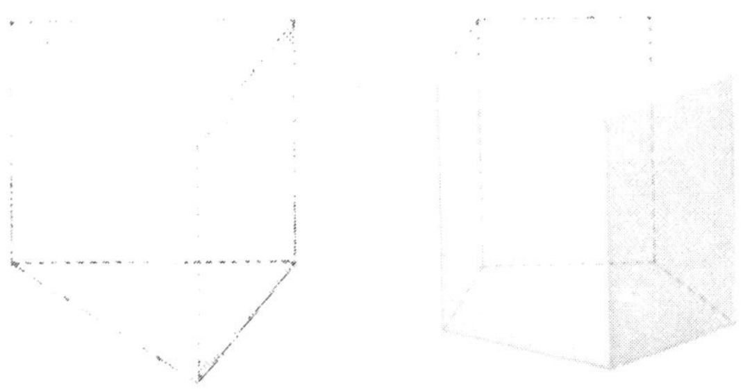

The ends of a prism may be any congruent figures: these may be triangles, quadrilaterals, or polygons of any number of sides. The diagram represents two prisms, one on a triangular base, the other on a pentagonal base.

Ex. 5. Draw the net of a triangular prism, whose ends are equilateral triangles on sides of 5 cm., and whose side-edges measure 7 cm.

(*Pyramids.*)

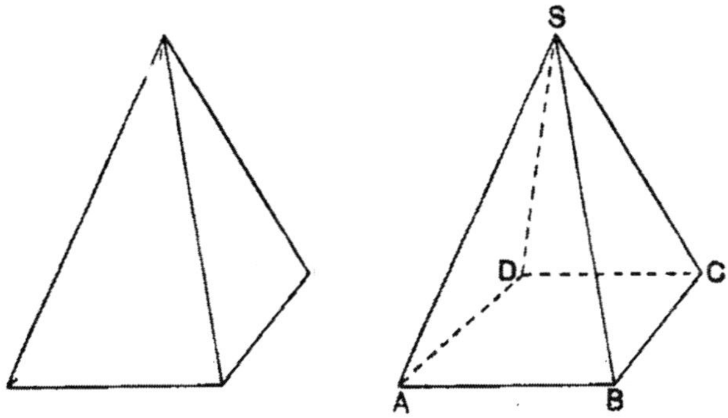

The solid represented in this diagram is called a **pyramid**.

The base of a pyramid (as of a prism) may have any number of sides, but the side-faces must be *triangles* whose vertices are at the same point.

The particular pyramid shewn in the Figure stands on a *square* base ABCD, and its side-edges SA, SB, SC, SD are all equal. In this case the side faces are equal isosceles triangles; and the pyramid is said to be *right*, for if the base is placed on a level table, then the vertex lies in an upright line through the mid-point of the base.

Ex. 6. Make a model of a right pyramid standing on a square base. Each edge of the base is to measure 3″, and each side-edge of the pyramid is to be 4″.

To make the necessary net, draw a square on a side of 3″. This will form the base of the pyramid. Then on the sides of this square draw isosceles triangles making the equal sides in each triangle 4″ long.

Explain why the process of folding about the dotted lines brings the four vertices together.

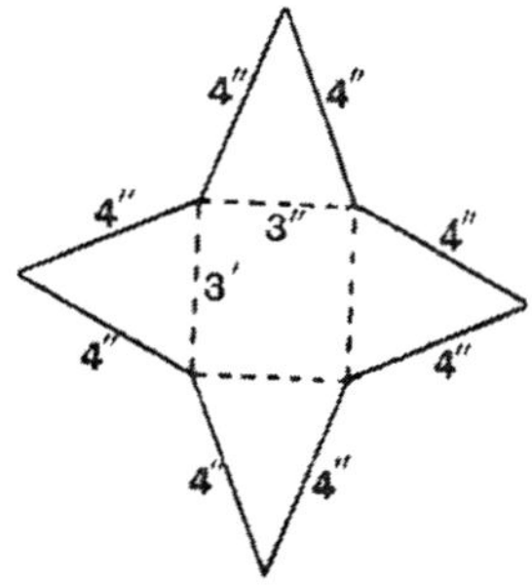

Another important form of pyramid has as base an equilateral triangle, and all the side edges are equal to the edges of the base.

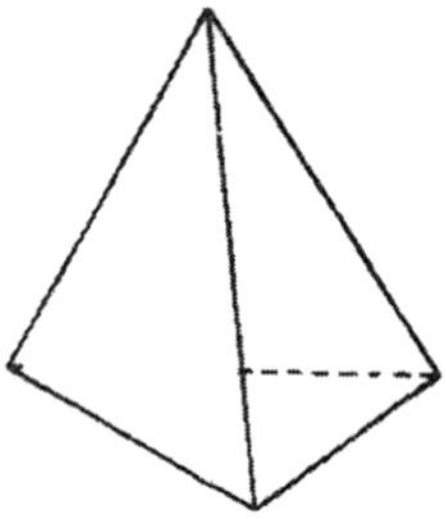

FIG. 1.

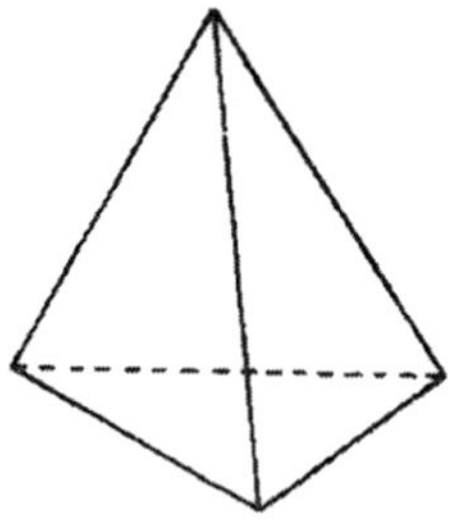

FIG. 2.

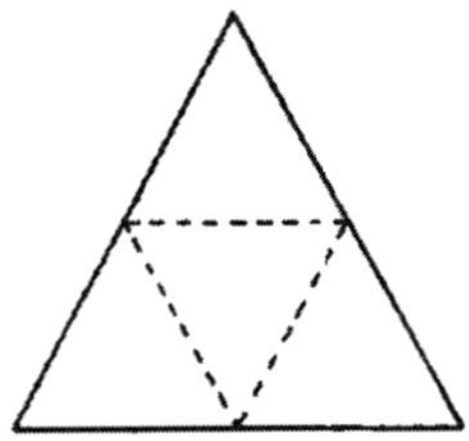

FIG. 3.

How many faces will such a pyramid have? How many edges? What sort of triangles will the side-faces be? Fig. 3 shews the net on a reduced scale.

A pyramid of this kind is called a regular **tetrahedron** (from Greek words meaning *four-faced*).

Ex. 7. Construct a model of a regular tetrahedron, each edge of which is 3″ long.

Ex. 8. What is the smallest number of *plane* faces that will enclose a space? What is the smallest number of *curved* surfaces that will enclose a space?

(*Cylinders.*)

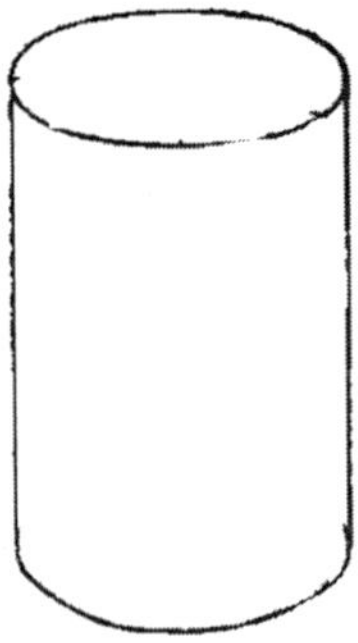

FIG. 1.

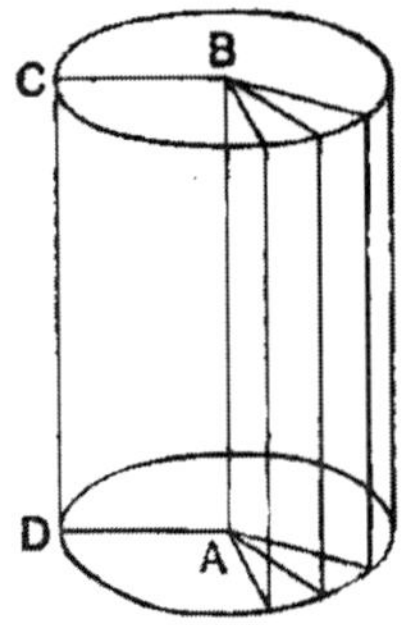

FIG. 2.

The solid figure here represented is called a **cylinder**.

On examining the model of which the last diagram is a drawing, you will notice that the two ends are *plane*, *circular*, *equal*, and *parallel*.

The side-surface is curved, but not curved in every direction; for it is evidently possible in one direction to rule *straight* lines on the surface: in *what* direction?

Let us take a rectangle ABCD (see Fig. 2), and suppose it to rotate about one side AB as a fixed axis.

What will BC and AD trace out, as they revolve about AB?

Observe that CD will move so as always to be parallel to the axis AB, and to pass round the curve traced out by D. As CD moves, it will generate (that is to say, *trace out*) a surface. What sort of surface?

We now see why in *one* direction, namely parallel to the axis AB, it is possible to rule *straight* lines on the *curved* surface of a cylinder.

It is easy to find a plane surface to represent the curved surface of a cylinder.

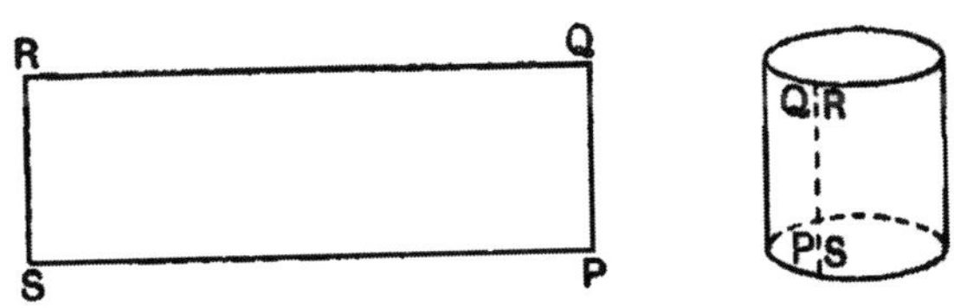

Cut a rectangular strip of paper, making the width PQ equal to the height of the cylinder. Wrap the paper round the cylinder, and carefully mark off the length PS that will make the paper go exactly once round. Cut off all that overlaps; and then unwrap the covering strip. You have now a rectangle representing the curved surface of the cylinder, and having the same area.

(*Cones.*)

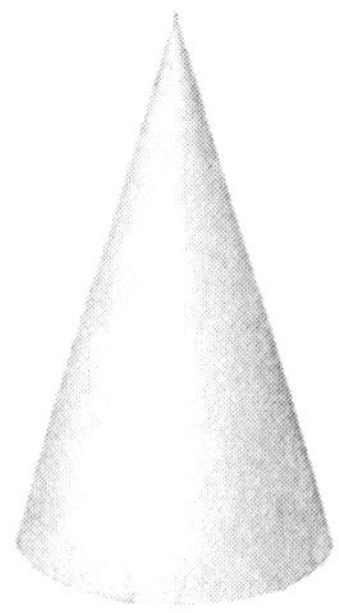

FIG. 1.

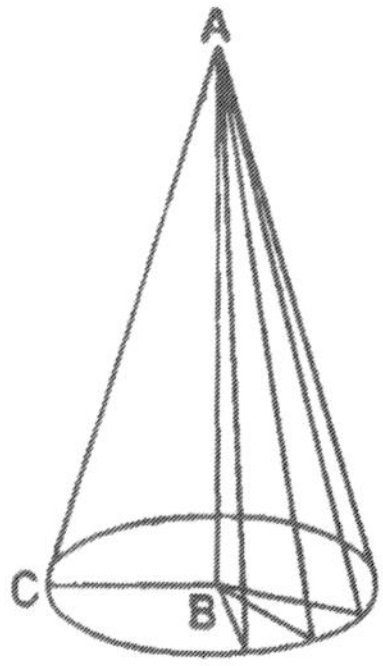

FIG. 2.

We have now to examine the model of a **cone**, of which a drawing is given above.

Its surface consists of two parts; first a *plane circular* base, then a *curved surface* which tapers from the circumference of the base to a point above it called the vertex. Thus the form of a cone suggests a pyramid standing on a circular instead of a rectilineal base.

Let us take a triangle ABC right-angled at B (Fig. 2), and suppose it to rotate about one side AB as a fixed axis. What will BC trace out as the triangle revolves? Notice that AC will always pass through the *fixed* point A, and move round the curve traced out by C. As AC moves, it will generate a surface. What sort of surface?

We now see that the kind of cone represented in the diagram is a solid generated by the revolution of a right-angled triangle about one side containing the right angle.

Ex. 9. Why must the △ ABC, rotating about AB, be *right-angled* at B, in order to generate a cone?

What would be generated by the revolution of an *obtuse*-angled triangle about one side forming the obtuse angle?

Ex. 10. What would be generated by an *oblique* parallelogram revolving about one side?

The curved surface of a cone may be represented by a plane figure thus:

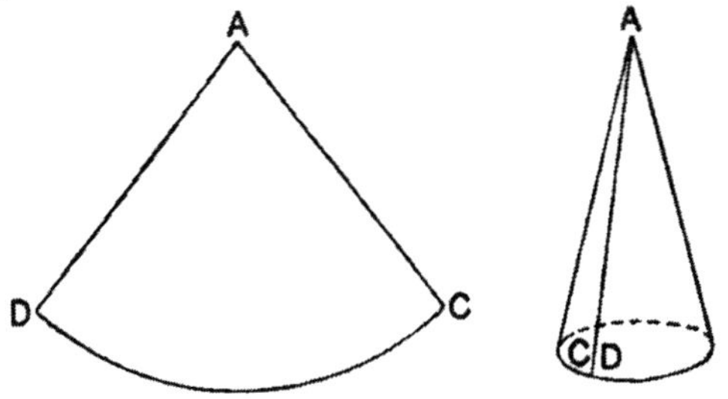

Taking the slant height AC of the cone as radius, draw a circle. Cut it out from your paper; call its centre A; and cut it along any radius AC. If you now place the centre of the circular paper at the vertex of the cone, you will find that you can wrap the paper round the cone without fold or crease. Mark off from the circumference of your paper the length CD that will go exactly once round the base of the cone; then cut through the radius AD. We have now a plane figure ACD (called a *sector of a circle*) which represents the curved surface of the cone, and has the same area.

(*Spheres.*)

The last solid we have to consider is the **sphere**, whose shape is that of a globe or billiard ball.

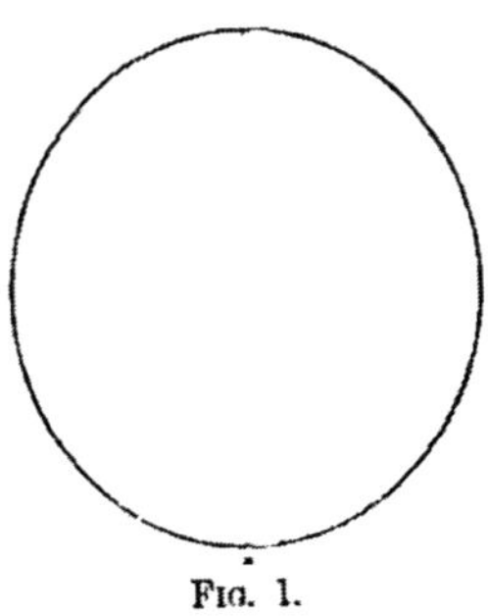

FIG. 1.

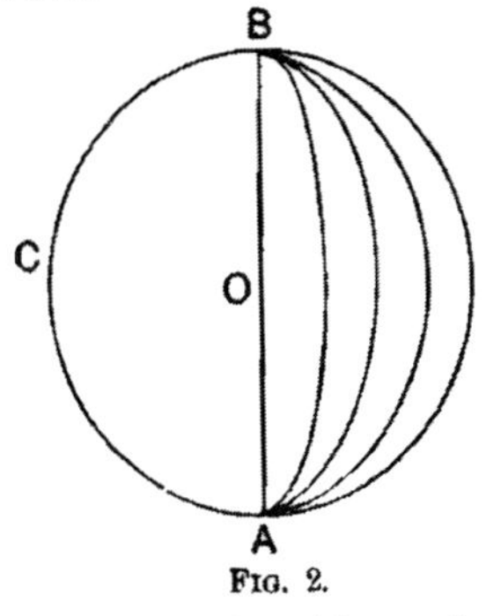

FIG. 2.

We shall realise its form more definitely, if we imagine a semi-circle ACB (Fig. 2) to rotate about its diameter as a fixed axis. Then, as the semi-circumference revolves, it generates the surface of a sphere.

Now since all points on the semi-circumference are in all positions at a constant distance from its centre O, we see that all points on the surface of a sphere are at a constant distance from a fixed point within it, namely the centre. This constant distance is the radius of the sphere. Thus all straight lines through the centre terminated both ways by the surface are equal: such lines are *diameters.*

Ex. 11. We have seen that on the curved surfaces of a cylinder and cone it is possible (in certain ways only) to rule *straight* lines. Is there any direction in which we can rule a straight line on the surface of a sphere?

Ex. 12. Again we have cut out a *plane* figure that could be wrapped round the *curved* surface of a cylinder without folding, creasing, or stretching. The same has been done for the curved surface of a cone. Can a flat piece of paper be wrapped about a sphere so as to fit all over the surface without creasing?

Ex. 13. Suppose you were to cut a sphere straight through the centre into two parts, in such a way that the new surfaces (made by cutting) are *plane*, these parts would be in every way alike. The parts into which a sphere is divided by a *plane central section* are called hemispheres. Of what shape is the line in which the plane surface meets the curved surface? If the section were *plane* but not *central*, can you tell what the meeting line of the two surfaces would be?

Ex. 14. If a cylinder were cut by a plane parallel to the base, of what shape would the new rim be?

Ex. 15. If a cone were cut by a plane parallel to the base, what would be the form of the section?

ANSWERS.

II. MEASUREMENT OF STRAIGHT LINES.

1. 1·8″; 3·2″. **2.** 4·5 cm.; 8·1 cm.
3. 1·8″; 1·3″; 3·1″. **4.** 8·5 cm.; 4·8 cm.; 3·7 cm.
5. 3·0″; 1·2″; 1·1″; 0·7″. **11.** 2·54 cm.
15. 400 m.; 560 m.; 80 m. **16.** 64 mi.; 4·3″.
17. 22 mi.; 11 mi.; 20 mi. **18.** 5 mi.
19. 36 ft. **20.** 29 ft. **21.** 17 ft.
22. 2½ mi. **23.** 31 mi. **24.** 50·5 m.

III. STRAIGHT LINES CONTINUED.

1. 2·54 cm.
4. AB = 1·3″ = 3·3 cm. CD = 2·2″ = 5·6 cm. EF = 0·8″ = 2·0 cm. GH = 3·0″ = 7·6 cm.
6. 2·83″; 2·12″; 1·41″; 0·70″.

IV. CIRCLES.

13. P is 2·5″ from A and from B. Q is 2″ from A and from B.
14. Two; one on each side of AB. **15.** Two. **16.** Two.
22. About 2·1″.

V. ANGLES.

7. 90°, 180°, 270°, 360°.
8. 30°, 150°, 216°. 8 min., 17 min., 1½ min.
9. 60°. **10.** 45°, 90°, 135°, 180°.
11. 30°, 60°, 90°, 120°, 150°, 180°. 60°, 60°.
23. (i) 115°; (ii) 40°; (iii) 27°.
25. (i) ∠ BOC = 37°, ∠ COA = 143°, ∠ AOD = 37°.
(ii) ∠ DOB = 151°, ∠ BOC = 29°, ∠ COA = 151°.
(iii) ∠ BOD = 137°, ∠ DOA = 43°, ∠ COB = 43°.
28. (i) 153°; (ii) 74°.

VII. DIRECTION. PARALLELS.

18. 7·8 km. **19.** 390 yds. **20.** 9·9 km. **21.** 10 mi.
22. About $30\frac{1}{2}$ mi. N. 25° W. **23.** 440 yds. **24.** 3 km.

VIII. PERPENDICULARS.

1. 6 cm. **5.** On the perpendicular bisector of AB. Two.
7. A square. **8.** 4·8 cm.
10. (*a*) A square. (*b*) A square. 90°. (*c*) 90°. **17.** 1·5″.

IX. TRIANGLES.

10. 45°. **11.** a=12 cm., B=35°, C=27°.
14. 50°. b=c=2·86″. (i) Isosceles, (ii) acute-angled.
15. 84°. a=5·5 cm., c=3·6 cm. **17.** 30°.

X. TRIANGLES CONTINUED.

14. ∠AXB=73°; ∠AXC=107°. **16.** 30°.
17. 30°. 12·5 cm. **27.** 430 yds. **28.** E. 37° N.
29. No, by about 0·1 of a mile. **30.** 134 ft.
31. 31°. **32.** 683 yds. **33.** 162 metres.
34. 505 ft. **35.** Due North. **36.** 159 yds.
37. Nearly 17 mi. **38.** 150 yds. **39.** 566 yds.; 400 yds.
40. About 32 mi. **41.** 132 ft. **42.** 20 ft.
43. 12 ft. **44.** Nearly 300 metres.

XI. QUADRILATERALS.

5. ∠ABC=58°, ∠BCD=122°, ∠ADC=58°.
7. The square and rectangle. **9.** 3·5″.
10. 2·1″. **11.** 5 cm. **14.** Five.
15. 60 metres. N.W. **16.** 19·1 cm. **17.** 175 yds.

XII. AREAS.

1. 180. **2.** (i) 200; (ii) 180; (iii) 200; (iv) 80.
4. The circle has an area of $78\frac{1}{2}$ squares; each of the other figures has an area of 80 squares.
5. 100. **6.** Four. **7.** (i) Three. (ii) Nine. (iii) Six.
9. 200 sq. ft. **10.** 1 sq. yd. 375 sq. yds.
11. (i) 180; (ii) 400; (iii) 900; (iv) 200; (v) 20.

12. 2″. **13.** Breadth = 4 cm. **14.** 1·6″.
15. 64 sq. ft. **16.** 100 sq. ft.; 2000 sq. ft.
17. 1200 sq. ft. **18.** 125 sq. mi. **19.** 625 sq. yds.
20. 40 yds.; 2000 sq. yds. **21.** 120 squares.
24. 5 sq. in. **25.** 36 sq. cm.
26. 1·53″; 2·06 sq. in. **27.** 28 sq. in.
28. 5·7 cm.; 45·6 sq. cm. **29.** 6 sq. in.; 3 sq. in.
31. 20 sq. cm. **35.** 4800 sq. yds.
36. Each = 1296 sq. m. **37.** 1·73 sq. in.
38. 20·3 sq. cm. **39.** 22·9 sq. cm.

XIII. MISCELLANEOUS CONSTRUCTIONS.

4. 5·0 cm. **5.** 2·0″. **6.** Each tangent = 9·6 cm.
7. A rhombus. **8.** Equilateral △. Square.
9. (i) 120°; (ii) 90°; (iii) 60°; (iv) 45°; (v) 36°.
12. 108°. **13.** 120°. **17.** 3·0″.

XIV. SOLID FIGURES.

1. The opposite faces ACED, BFGH are squares.
2. Each face is a square. **8.** Four. Two.
9. A cone with a conical cavity at one end.
10. A cylinder with a conical cavity at one end and a conical peak at the other.
11. No. **12.** No. **13.** A circle. A circle.
14. A circle. A circle.